Knowledge Management and Organizational Learning

Volume 2

Series editors
Ettore Bolisani, Padova, Italy
Meliha Handzic, Istanbul, Turkey

More information about this series at http://www.springer.com/series/11850

András Gábor • Andrea Kő

Editors

Corporate Knowledge Discovery and Organizational Learning

The Role, Importance, and Application of Semantic Business Process Management

Editors
András Gábor
Corvinno Ltd.
Budapest, Hungary

Andrea Kő
Corvinus University of Budapest
Budapest, Hungary

ISSN 2199-8663   ISSN 2199-8671   (electronic)
Knowledge Management and Organizational Learning
ISBN 978-3-319-80456-9   ISBN 978-3-319-28917-5   (eBook)
DOI 10.1007/978-3-319-28917-5

Printed on acid-free paper

This Springer imprint is published by Springer Nature
The registered company is Springer International Publishing AG Switzerland

Prologue: Book Series on "Knowledge Management and Organizational Learning"

We are happy to announce the publication of the second Volume of the Book Series on "Knowledge Management and Organizational Learning" that was launched in 2015 with the factive contribution of our colleagues and friends of the International Association for Knowledge Management (IAKM). The Book Series recalls the main mission of our Association: to support the development of Knowledge Management (KM) as a scientific discipline. Despite its increasing importance in academia and in practice, KM still suffers, like any other "new area," from a problem of "foundation." It aims to become an independent field, but as it has multidisciplinary roots—from psychology to computer science, from organizational science to business administration, just to mention some—it requires an integration of different perspectives and a robust clarification of its conceptual references. Research and practice often branch off in multiple directions, and no clear consensus on concepts and methods has emerged so far.

As scientists and professionals involved in KM, we need to develop "core" theories, common approaches, and standard languages that can help us see the problem of managing knowledge under the same shared perspective. We also need to explore emerging new interdisciplinary and transdisciplinary ideas and align them with the foundation. The way to reach a credible agreement on what we are doing and to set a common ground for our future work calls for a capability to discuss, exchange, and, maybe, contrast our ideas and positions freely and openly. We need a place where we can do this in a rigorous but, at the same time, friendly atmosphere.

This Book Series is an integral part of this mission. What inspires it is not the acceptance to a particular "school of thought" or "ideological" position, as sometimes happens even in the scientific world. Rather, what inspires it is a vision of KM as a "playground" where there is a lot to research, discover, and innovate and; where curiosity, dialogue, and openness to disagreements are the key ingredients.

With the same scrupulousness of scientific publications, but with a broader scope and more relaxed constraints than those that may characterize other editorial channels, the Series puts an emphasis on free discussions of new theories, methods, and approaches, on visions of the future and advances in the field; on critical reviews of recent or past empirical evidence, and on formulating ideas for new practical methods or applications. It aims to offer a constantly updated reference to researchers, practitioners, and also students involved in the field of KM and its application.

The first inaugural Volume, *Advances in Knowledge Management: Celebrating Twenty Years of Research and Practice* (edited by Ettore Bolisani and Meliha Handzic), has the goal to assess the "state of KM" as a discipline, the points that still have to be developed, and the directions that the research and practice of KM can or should take. Subdivided into three parts (emblematically entitled "Analyzing the Past," "Acting in Present," and "Predicting the Future"), the book collects nine authoritative chapters that examine KM from different viewpoints and by considering diverse positions and aspects. It analyzes not only the research but also the practice in this field and provides a picture that, for sure, will be stimulating for readers. It makes a good starting point for our Series.

The second Volume, *Corporate Knowledge Discovery and Organizational Learning* (edited by András Gábor and Andrea Kő), has a more specific and applicative content. It summarizes the results of a big research project—*ProKEX*—involving researchers of different Institutions and with a leading role of the Corvinus University of Budapest. The ProKEX research addresses a peculiar issue of organizational knowledge management, i.e., is it possible to design a software application that helps companies to automatically trace, store, and deliver all the key details of its internal processes? In a world where the efficiency of organizational processes is vital, but, at the same time, there is a demand for their continuous updating, it is important for companies to keep track of, represent, and memorize their internal routines in a way that can be easily retrieved and made available to employees and executives. The book provides an interesting perspective on how it is possible to extract, organize, share, and preserve the knowledge embedded in organizational processes in order to enrich the organizational memory in a systematic and controlled way, to support employees to easily acquire their job role-specific knowledge, and to help govern and plan the investments in human capital. The various chapters, written by different researchers, not only discuss the conceptual foundations of the project but also describe the implementation details of the software applications that the team has developed and tested in real-life situation. Both specialized readers and, more generally, people interested in advanced KM issues will enjoy the book.

International Association for Knowledge Management
www.iakm.net

Padova, Italy   Ettore Bolisani
Sarajevo, Bosnia and Herzegovina   Meliha Handzic

Preface

Due to rapid developments in technology and in socio-economic transformation, the management of organizational knowledge is becoming increasingly important. This knowledge resides in many places, which includes organizational processes. The extraction of hidden knowledge from business processes, their articulation, and their transfer is a major challenge. Knowledge management and practices combined with semantic business process management open up many new opportunities in the management of organizational knowledge. The prospects are even more challenging if we consider the new approaches of data science since the volume of digital content is growing rapidly, and in many cases traditional methods are no longer suitable. By nature, digital content is diverse, making it another important driver to investigate new methods and procedures.

This book deals with the following key questions:

- How can we utilize the embedded knowledge of business processes in organizational knowledge management and in employees' training?
- How can we connect business processes to organizational knowledge bases, assuming that changes in business logic and business processes would be reflected in the knowledge base?
- How can we make sure that the knowledge required through business processes will be attainable and transferable to employees assigned as performers to specific job roles?

In this book, the authors will discuss the role, importance, and applicable methods of semantic business process management in the context of organizational knowledge management. The ProKEX project provided the right framework for discussion.

The aim of the ProKEX project was to develop a complex methodology and application that addresses organizational knowledge management in terms of knowledge elicitation, knowledge representation, and knowledge sharing. The goal of the ProKEX solution is to extract, organize, share, and preserve knowledge embedded in organizational processes in order to (1) enrich organizational knowledge bases in a systematic and controlled way, (2) support employees to be better able to acquire their job role-specific knowledge, (3) and help govern and plan human capital investment. In order to keep the methodology consistent, the business process management (BPM) approach and organizational knowledge management (KM) approach were combined. BPM is supposed to be a fairly standard methodology and is the most appropriate for business logic and hence is an outstanding starting point to capture organizational knowledge (i.e. embedded into processes). KM provides ample procedures to represent and elaborate knowledge.

The book is divided into seven chapters which detail the ProKEX approach. Each chapter highlights one specific part of the methodology and illustrates the contribution to the overall approach and the impact within the framework of organizational knowledge management.

The first chapter provides an overview in a nutshell of the ProKEX research and its theoretical background related to knowledge management and business process management. This chapter creates the framework for the chapters that follow, which provide a more detailed discussion of the ProKEX components.

The second chapter illustrates the semantic business process management and discusses these aspects of the ProKEX solution. The method of extracting, organizing, and preserving embedded knowledge is explained in detail. Furthermore, it introduces a method how to translate the business process model into process ontology.

The third chapter presents a text mining solution, which has a key role in the knowledge discovery of the business processes. ProMine is a text mining application used for ontology enrichment based on the extraction of deep representations from business processes. ProMine extracts new domain-related concepts using a new filtering mechanism to filter the most relevant concepts, based on a novel hybrid similarity measure.

Characterizing, structuring, and systematizing the knowledge assets of an organization are a major challenge. The fourth chapter describes how the integrated STUDIO knowledge-based system supports organizations in applying and evaluating knowledge, in a guided learning process. It empowers employees to adapt changes to their own context quickly and supports the conversion of organizational learning into action.

The fifth chapter details the ProKEX solution in a technological context. The main focus in this chapter is devoted to the bridging process model, especially job roles, with the related domain knowledge. The main technological components of ProKEX is therefore ontology tailoring.

The sixth chapter fosters the vision of a context-aware and context-rich assessment for self-assessment. It describes a new adaptive test, which is flexible in terms of the knowledge to assess and adaptive in terms of the knowledge of an individual worker. The concept details the STUDIO knowledge-based system in terms of adaptive testing and knowledge exploration.

The seventh chapter introduces perspectives and the further development of the ProKEX solution and shows how the ProKEX approach may help the process owner to improve the business processes under their control. It provides a brief overview of the theoretical background of the necessary ontology matching procedures and presents a concept which integrates several methodological results of the ProKEX project—such as XSLT conversion, text mining, similarity measures, and ontology tailoring. This book may be of use to both practitioners and researchers. Since the separate chapters cover quite a wide array of business process management and knowledge management, including the emerging data science, we believe that each chapter can contribute to the exploration of one or more specific scientific and practical areas. The main areas addressed in the book cover knowledge engineering, knowledge discovery, semantic web technologies, ontology management, knowledge transfer, and organizational learning.

The authors wish to express their gratitude to Ettore Bolisani, the volume editor, and the International Association for Knowledge Management (IAKM), who made it possible to publish research results, while special thanks go to Dr. Ljiljiana Stojanovic and Dr. Ioana Ciuciu for their valuable comments.

Budapest, Hungary   Andrea Kő
November, 2015   András Gábor

Contents

Corporate Knowledge Discovery and Organizational Learning: The Role, Importance, and Application of Semantic Business Process Management—The ProKEX Case

András Gábor, Andrea Kő, Zoltán Szabó, and Péter Fehér

1 Introduction: From BPM to SBPM

Business process management (BPM) focuses on business operations, as well as key value adding and supporting activities of organizations. BPM integrates several methods and techniques for modelling, analysing, reorganizing, operating and monitoring the processes of an organization (Scheer et al. 2002). Business process management originates from global business trends as a management method to facilitate strategic alignment by streamlining business processes, and harmonizing organization and technology. Strategic alignment was originally defined as concerning the inherently dynamic fit between external and internal domains, such as product/market, strategy, administrative structures, business processes and IT (Henderson and Venkatraman 1993). It is argued that economic performance is enhanced when the right fit between external positioning and internal arrangements is found.

Although BPM is considered as a strategic tool of business revitalization, its popular interpretation focuses on the modelling and implementation aspects: BPM is about describing business processes in a complex modelling tool and implementing the process in supporting applications (ERP, workflows). The emphasis is on the effective use of models for automatic generation of IT applications. BPM has also integrated many quality management related approaches (Lean, six sigma, maturity models), controlling and strategic

A. Gábor (✉)
Corvinno Technology Transfer Centre, Budapest, Hungary
e-mail: agabor@corvinno.com

A. Kő • Z. Szabó • P. Fehér
Corvinus University of Budapest, Budapest, Hungary
e-mail: andrea.ko@uni-corvinus.hu; zoltan.szabo@uni-corvinus.hu; peter.feher@uni-corvinus.hu

© Springer International Publishing Switzerland 2016
A. Gábor, A. Kő (eds.), *Corporate Knowledge Discovery and Organizational Learning*, Knowledge Management and Organizational Learning 2,
DOI 10.1007/978-3-319-28917-5_1

management approaches (balanced scorecard), to facilitate the sophisticated utilization of the concept.

Strategic alignment is a dynamic process: continuous adjustment of strategy, organizational structure, technology platform and skills (knowledge) is a key issue in today's business environment, and more important than ever. Frequent changes in the environment (regulation, requirements of compliance, changing user needs, shorter product life-cycles, customization, emerging new technologies and "superconductivity" of markets) make harmonization between processes, skills, human resources and technology a challenging task. BPM is traditionally an effective tool for revitalizing out-dated business processes, and for increasing their productivity and improving quality. As a holistic approach BPM is appropriate for the analysis, improvement and control of business processes. To support the dynamic nature of strategic alignment BPM related activities can be organized into a systematic life-cycle.

The cyclic approach of BPM involves the following phases (based on Scheer et al. 2002; Anonymous 2005):

- A business process strategy that defines the strategic goals and creates a process portfolio.
- Process documentation that creates the process models and gathers relevant information.
- Process analysis and design that investigates process-related problems (cycle time, cost, quality, etc.), and optimizes the process, defining an integrated system of processes, organization and technology.
- Implementation and change management that ensures the realization of plans, IT projects and organizational changes.
- A process operation that maintains an appropriate organizational environment for the utilization of processes.
- Process controlling/monitoring that gathers process-related KPIs and provides a feedback mechanism for further development.

This cycle provides a comprehensive system that enables organizations to improve their process maturity and also to use the process management concept as an effective tool of strategic alignment. The practice puts more emphasis on process modelling that supports the design and implementation of IT applications. The maintenance and the systematic integration of these models is a major burden, and for many organizations the obsolescence of the models is a recurring issue. This problem indicates that BPM is not just a favourite toy of the IT units and consultants, but a continuous effort to describe organizational knowledge about operations in the form of models, model based solutions, measurement and controlling methods, and organizational arrangements (roles, responsibilities, etc.). In this respect BPM is a form of organizational learning: about strategy, organizational structure, IT and the knowledge necessary for operations. As we pointed out, each phase of the BPM life cycle is knowledge dependent and should be supported by knowledge management methods and tools (Gábor and Szabó 2013).

2 Challenges

> The time will come, when the knowledge created from a business process will be more important than the execution of the actual business process itself[1]

This section details those problems and challenges that we target in our research. It is almost a trivial fact in the modern economy that the regulatory, social and economic environment is fairly complex and they are continuously changing at both a local and global level. One of the consequences of this for organisations is the growing demand to efficiently manage corporate assets as well as their intellectual capital.

Intellectual capital consists of many components; one of the main components is the task knowledge needed to perform different tasks. The knowledge that needs to be updated is not general but rather very much focused on organisational requirements, and taking a closer look it is the knowledge that is embedded in processes for which employees are responsible. The problem that management must deal with is to determine what the embedded knowledge is and how to *extract it from the various processes*? This is a dynamic challenge, and in a sense the most frequently changing component is the process, thus if there is any change, it will initially occur in the processes.

There is usually a bottleneck in the maintenance of the process models, and in many cases the explicit process models do not cover the actual processes due to dynamic changes, obsolescence, and lack of maintenance. Updating of processes can be triggered for many reasons. The following are by far not an exhaustive list, but only examples: there is a need for compliance checking (e.g. information security, data privacy regulatory measures); technology changes (e.g. many companies introduce some elements of ERP in mobile technology); there is change in the organisational setup and the position-job role-task assignment is changing. It is in the vital interests of management to keep the process models updated, and if an automated or semi-automated procedure is available to carry out maintenance it would constitute positive progress in the business process management. The question is how, when and who should suggest the need to update? One opportunity could be an analysis of knowledge levels and knowledge gaps on the part of the performers that could then indicate the need for changes as well as updates in the process models.

---

[1] El Sawy, Omar A., and Robert A. Josefek Jr. "Business process as nexus of knowledge." in Holsapple, Clyde, ed *Handbook on Knowledge Management 1: Knowledge matters*. Vol. 1, pp. 425–436. Springer Science & Business Media, 2013.

Another aspect of the problem regarding the preservation of intellectual capital: employee turnover is greater than in the past, so companies need appropriate methods for the codification of their "human capital" in order to transfer it to the new employees. This is especially the case when an organisation is geographically dispersed but there are strict regulations to identically perform the same job under a strict technological and quality regime. While the management of intellectual capital generally concentrates on the necessity of knowledge elicitation, here the focus is rather on *knowledge transfer*.

Due to the nature of the everyday operation of most companies, including organisations both in the competitive sector and in administration and non-profit sectors, the aforementioned knowledge resides in the heads of the human performers; hence job-related knowledge is better addressed than purely task-related ones. In conclusion, the accelerated knowledge renewal cycle affects human resources, in other words job-related knowledge must be renewed regularly. Many questions arise in regard to what form and to what extent knowledge must be renewed, since it depends not only on the external circumstances, but also on the previous knowledge of every employee, on the organisational setup, on the position-job role(s) assignment in the organisation, on the learning profile, as well as learning behaviour. Consequently, knowledge transfer cannot be efficient, nor successful if a relatively adequate map of knowledge gaps is not available. Knowledge gaps are very useful to orient where and what to learn additionally. The question here is what are the results of *an adequate map of knowledge gaps*?

BPM phases and activities are dependent on organizational knowledge; BPM can be aligned to knowledge management. Figure 1 presents the link between knowledge management life cycle phases and business process management fields. The external cycle details the phases in the knowledge management life cycle, while the internal cycle deals with process management life cycle phases (Gábor and Szabó 2013). State of the art is discussed in the next chapter according to the figure below.

The initial phases of the Business Process Management Life Cycle (BPMLC), process strategy and documentation/modelling are highly dependent upon Knowledge Management: knowledge discovery and codification are key enablers. Management and the project participants should understand business strategy and interpret it in relation to the process. Market and operative knowledge should be combined and hidden activities of the organization should be explored and documented in an explicit form.

Process documentation may be a main tool of knowledge transfer during the analysis of the process: it provides a common platform and the knowledge base that facilitates the understanding, diagnosis and redesign of the process. Knowledge of several actors and stakeholders should be integrated into a model base to facilitate further work, such as process improvement. Documentation (in the form of complex models) may also have a major role in the training of new employees.

The design phase requires intensive knowledge sharing to enable cooperation and cross-functional teamwork, to facilitate the integration of different aspects of

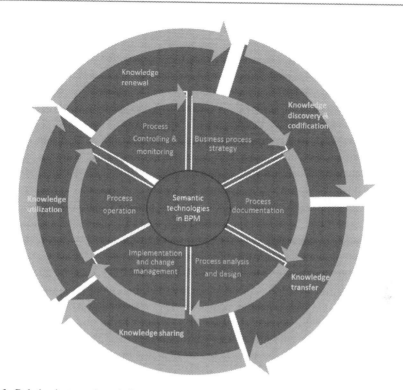

Fig. 1 Relation between knowledge management life cycle phases and business process management fields (Gábor and Szabó 2013)

process relevant activities and to support innovation (generation of new ideas for process redesign).

Implementation and change management is a risky and complex phase, in this key period of a project intensive organizational learning is necessary. Training, involvement of the relevant interest groups, and re-establishing a new common platform of operations requires knowledge sharing.

The operation phase traditionally emphasizes the organizational aspects (process managers, sponsors, specialists) that enable sustainability and partly maintainability of process-oriented operations. Knowledge utilization is a key issue in this phase, knowledge is the major facilitator of process execution, so it should be explicitly documented, and be available for the relevant participants for internalization. The maintenance of this operation-related knowledge can be an integral part of the maintenance of process models. To ensure up-to-date process know-how, knowledge aspects of process-related activities must be identified and codified.

Monitoring phases are strongly related to knowledge renewal. The cyclic nature of BPM requires systematic evaluation, and adaptability to integrate new knowledge, and to renew the whole BPM system by providing new and relevant knowledge as an input for the strategic phase.

We argue that knowledge-related activities are key elements of successful process management for dealing with complexity and turbulence. Focused management of process knowledge is necessary in organizations. Semantic technologies, as facilitators of transforming process models into executable applications are frequently discussed in literature (e.g. Hepp and Roman 2007; Davies et al. 2009; Warren et al. 2011), in this book we will extend this to the complete BPM lifecycle, introducing a prototype that provides comprehensive support for process management. Using semantic technologies, knowledge management tools can be implemented to facilitate the management of process and job related knowledge elements, enabling customized training programs and the efficient maintenance of knowledge.

3 State of the Art

Knowledge is a key resource of companies and a critical factor in their competitiveness and economic growth. This section deals with the theoretical foundations of our research; we focus on knowledge discovery and creation in terms of organizational learning and knowledge sharing in respect to intellectual capital management. Business process management areas, such as the business process management life cycle, its knowledge related challenges and SBPM are also introduced.

3.1 Intellectual Capital

Intellectual capital has its origin in industry and consultancy. Knowledge assets or intellectual capital are treated as the raw materials of a value creation process of the organization that help them to create and refresh organizational competencies over time (Kaplan and Norton 2004; Marr et al. 2004). Knowledge process capabilities are the abilities of an organization to utilize knowledge assets to generate valuable knowledge through a series of managerial processes (Lee and Choi 2003; Tanriverdi 2005).

Intellectual capital has a long history going back to 1969 with Tobin's q ratio (Tobin 1969), developed by the economist James Tobin. This is defined as the ratio of the stock market value of the firm divided by the replacement cost of its assets.

The term became popular worldwide from the 80s, when Kaplan and Norton introduced the concept of the "balanced scorecard" (Kaplan and Norton 1992). Their book *"The Balanced Scorecard: Translating Strategy into Action"* is a decisive resource in the field. There are several other important milestones in the history of intellectual capital, such as in 1989 when "Invisible Balance Sheet" was published by Sveiby (1989), in 1990 when Leif Edvinsson was appointed 'Director of Intellectual Capital' at Skandia AFS; and in 1997 when "Calculated Intangible Value" was introduced by Stewart (1997).

The term has several definitions; the most common one is that intellectual capital is the composition of *human capital, customer capital and structural capital* (Petrash 1996). *Human capital* is embedded in employees' heads and refers to people's knowledge, skills, capabilities, work-related competence, experience and expertise. It is typically tacit knowledge. *Customer capital* is the value of relationships with customers, suppliers and allies (Stewart 1997). A common form is customer loyalty.

Structural capital consists of tangible elements within the organization, such as organizational routines, organizational structure, management processes and corporate culture (Lee and Choi 2003; Gold and Arvind Malhotra 2001). Such elements remain in the company after employees go home at night.

Intellectual capital and knowledge management were developed in parallel. Both fields have a similar goal. They both have the aim to understand the role of knowledge and its management in companies' success and competitiveness (e.g., Nonaka and Takeuchi 1995). Hsu, I-Chieh and Sabherwal, Rajiv investigated the interrelations of intellectual capital and knowledge management through empirical methods using data from 533 companies in Tajwan (Hsu and Sabherwal 2012). They argued that intellectual capital and knowledge management affect performance indirectly through dynamic capabilities, innovation, and efficiency. Based on their data analysis they concluded that intellectual capital facilitates knowledge management and dynamic capabilities; knowledge management, a learning culture, and dynamic capabilities facilitate innovation; and innovation and efficiency facilitate performance. However their empirical investigation presented some unexpected results, e.g. neither intellectual capital nor knowledge management affects efficiency; and dynamic capabilities do not directly affect firm performance.

Intellectual capital became a decisive factor at the end of 80s. The most important reason for its popularity is that knowledge became a critical factor in organizational performance and economic growth. Value-added and customized services were required in the field of software development, consultancy, insurance and finance, and distinguished these companies from each other based on the built-in knowledge. The market value of Microsoft e.g. in 1997 was equal to the total market value of Boeing, McDonalds, Texaco, Time-Warner and Anheuser Busch (Skyrme 1999). The question was obvious: is Microsoft overpriced or are the other companies underpriced? How can we explain "market value"? What components determine it? These questions led to an increasing need for the measurement of intellectual capital, while traditional financial reports were insufficient to answer these questions. Understanding the nature of intellectual capital in an organization supports the leveraging of its knowledge assets.

3.2 Measurement of Intellectual Capital

There is a plethora of literature on the measurement of intellectual capital, so we will cite only some decisive approaches here. In 1993 Leif Edvinsson wrote about the hidden intellectual assets of Skandia AFS as a supplement of the annual report that was the "Skandia Navigator". It was the first time that the term "intellectual capital" had been used.

In their approach intellectual capital is measured through the analysis of up to 164 metric measures (91 intellectually based and 73 traditional metrics) that cover five components: (1) financial; (2) customer; (3) process; (4) renewal and development; and (5) human.

The Balanced Scorecard was developed by Robert Kaplan (Harvard Business School) and David Norton as a performance measurement framework (Kaplan and Norton 1996). It is used all over the world for several strategic management-related purposes: to translate strategy for operational actions, to improve communications both internal and external, and to monitor organization performance against strategic goals. It is a strategic planning and management system that has non-financial (knowledge related) performance measures to complement traditional financial metrics. This extension gives managers and executives a more 'balanced' view of organizational performance. A company's performance is measured by indicators covering four major focus perspectives in the first version of the Balanced Scorecard (BSC): (1) financial perspective; (2) customer perspective; (3) internal process perspective; and (4) learning perspective. These indicators are based on the strategic objectives of the firm. Learning perspective has a direct relation to intellectual capital. The model has another important message for the organizations through the strategy map; to achieve better performance the company has to invest in knowledge assets. Sveiby and his colleagues from the Konrad Group explained the difference between the stock market value of a firm and its net book value through the three main categories of intellectual capital; human capital, structural capital and customer (or relational) capital (Sveiby 1989; Sydler et al. 2014). These three categories have become a de facto standard.

3.3   Organizational Learning as Grabbing Organizational Knowledge

Organizational knowledge creation is one of the key areas investigated by knowledge management. The most important issue is how to utilize organizational learning to become a "learning organization", which can facilitate organizational survival and growth. The creation of organizational knowledge is usually discussed in the literature based on the research results of organizational learning. Organizational learning is an interdisciplinary field. The related disciplines are management science, psychology, strategy, sociology and cultural anthropology (Easterby-Smith 1997). In spite of the fact that there is no single framework used to describe it, there is a wealth of literature on this theme.

Discussion in the literature is extensive (Easterby-Smith and Lyles 2011; Jain and Moreno 2015), on how knowledge management and organizational learning are connected. In our context organizational learning is interpreted in the life-cycle of knowledge management as a tool to support the continuous development of organizational knowledge, and to enhance organizational processes.

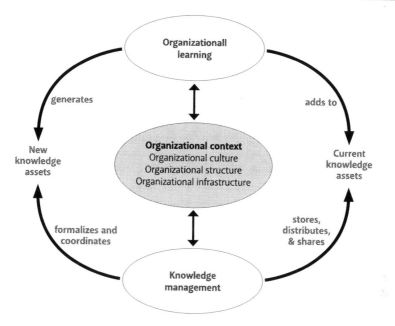

Fig. 2 The organizational learning and knowledge management environment (Pemberton and Stonehouse 2000)

Organizations have to continuously adapt to a global and continuously changing environment, in which organizational learning has a role to continuously create and internalize new knowledge and/or reshape the existing one. Pemberton and Stonehouse (2000) emphasize the importance of formalizing new knowledge, and coordinating the knowledge creating process, while storing, distributing and sharing existing knowledge. That is how the concepts of organizational learning and knowledge management are connected (Fig. 2).

Organizational learning is one of the approaches for knowledge management, including knowledge discovery and formal representation. Argyris and Schön (1978) and Argyris (1992) intensively analysed the phenomena of organizational learning. Argyris and Schön define organizational learning as a reactive action, and as the "detection and correction of error", while Fiol and Lyles (1985) emphasize the importance of "improving actions through better knowledge and understanding".

Based on the previous works of Argyris and Schön (1978), Pawlowsky (1992), Child et al. (1994) and Child et al. (2005) have identified three levels of organizational learning and completed these levels with pragmatic interpretations (Table 1).

Applying these loops means completing the knowledge management lifecycle that results in changes in the organizational processes. In our model these feedbacks (completing the loops) are supported by technological solutions in order to support the efficiency of organizational learning. Senge (1992) extends the previously discussed three levels by an understanding of the activity of the firm that initiates

Table. 1 Levels of organizational learning (Child et al. 2005, p. 273)

Levels	Theoretical approach	Pragmatic approach
High	*Learning—"deutero learning"* Learning how to learn so as to improve the quality of the organizational learning process itself	*Strategic learning* Changes in managerial mindsets, especially in understanding the criteria and conditions for organizational success
Middle	*Reframing—"double loop"* Changes of existing organizational frameworks. Involves questioning existing systems. Oriented towards survival in changing environmental conditions	*Systematic learning* Changes in organizational systems, with an emphasis on learning how to achieve better integration of organizational activities
Low	*Routine—"single loop"* Improvements and adjustments to optimize performance within the limits of existing organizational frameworks and systems	*Technical learning* The acquisition of new specific techniques such as more advanced production scheduling, or managerial techniques such as more advanced selection tests

organizational learning. In this approach adaptive learning organizations modify themselves continuously to match the changing requirements of the environment, while creating new skills and seizing new business opportunities through generative learning organizations. Although we are discussing organizational learning it is hard to separate the learning process of the organization from the members of the organization. Learning is achieved through internalizing organizational experience in personal behaviour; therefore it is important to receive personal feedback in the knowledge management lifecycle.

Dixon (2000) emphasizes the importance of exploiting existing experience through understanding it, and creating common knowledge inside a group. Creating common knowledge requires connecting the performed activities and their results, and deciding whether a team should change their approach. The process should not stop at this point. Either the experience is documented in a formal way (explicit knowledge) or resides only in the heads of the employees (tacit) a knowledge transfer system is required in order to leverage common knowledge. The receiving team or individual adapts the knowledge for use in practical contexts that can also be interpreted as a learning process.

Organizational learning processes could also concentrate on acquiring and using external knowledge (March 1991; Hedlund 1994; Volberda 1996), but the existing knowledge and capabilities of both at an organizational and individual level limit the absorption capacity of the organization. Existing background knowledge (learning capability) helps to identify and realise the value of external knowledge (Cohen and Levinthal 1990), but it is also necessary to learn skills (see the concept of deutero learning). Individuals and organizations adapt new knowledge more easily if it is closer to their existing active knowledge base. Organizational process models reflect this already existing knowledge through the granularity of the models: with wider background knowledge, models are simpler, and with less knowledge models need to be more detailed providing more help for the actors.

Based on the required absorptive capacity of organizations Gavin et al. (2008) analysed the sustainability and development opportunities of organizational learning processes and identified three main building blocks that organizations should understand: a supporting learning environment, a concrete learning process and the role of leadership.

A supportive learning environment, in which employees have the opportunity to question existing practices, are encouraged to disagree on a solid basis, and even to make mistakes and present minority viewpoints. This environment keeps the environment and the organization a dynamic one in which new ideas and solutions can be born. It also provides valuable feedback on existing practices and enables the development of organizational processes.

In our approach this perspective provides actors with the full context of their activities: if the actors understand not only the processes and the rules, but why these rules should be applied and how the processes are organized, they will be more capable of performing their tasks, resulting in more efficient organizational work.

While the environment creates a supporting and accepting culture, **concrete learning processes** make it possible to consciously exploit the learning opportunities an organization has. Therefore companies should apply formal knowledge management processes targeting both internal (experiences, problems to solve, employees' skill development) and external knowledge (competitive intelligence, customer behaviour, technology trends). The lifecycle approach, and the system architecture support the above mentioned knowledge management procedures (see Chapter "Ontology tailoring for job role knowledge" for more details).

Despite accepting the importance of learning organisations, Örtenblad (2015) emphasized the importance of conscious learning processes that fit the requirements and context of their environment: they should consciously select the strategic knowledge area, and consider how to develop their learning organization. In order to keep the learning processes alive, leadership should demonstrate support that reinforces learning, and demonstrate the value of the supportive culture through actively involving themselves in the processes.

3.4   Semantic Business Process Management

Business Process Excellence is the generally accepted major goal of process management (Scheer et al. 2002). Key dimensions of a process—time, cost and quality—are always at the focal point of business initiatives, while comprehensive process management can be a strategic asset for the company. BPM also represents a perfect tool for efficiently supporting the organization's day-to-day operation: regulations, roles and responsibilities are clearly defined in models that can be interpreted in an easy-to-use form for the relevant staff. Process oriented measurement—monitoring of process performance and reporting of process KPIs—is a common practice that enables the smooth operation of many huge organizations.

Maddern et al. (2014) discusses the importance of a holistic approach as well as the end-to-end process management, and discussed BPM related symptoms of fragmentation in modelling and optimization. They reported that the on-going maintenance of a process infrastructure is a very challenging task for organizations. End-to-end process management raises the question of complexity, especially in the case of inter-organizational processes.

The key dimensions of an organization (structure, IT applications, process, knowledge and actual practice) are always divergent leading to organizational misalignment, and this continuous obsolescence of documented, explicit organizational knowledge can be the root cause of many deviations and problems in productivity. Although the life cycle based approach of BPM can be a powerful tool for strategic alignment, in reality the systematic application of each phase requires concerted effort and supporting tools. The knowledge content of processes is usually a snapshot collected during the process modelling campaigns, but the changes in the market, customer needs, regulation, and products, etc. will always challenge the explicit knowledge embedded in the processes. Knowledge is distributed in hundreds of processes and may be represented in a heterogeneous form. To facilitate the maintenance and widespread utilization of the knowledge base the established models should be appropriate for automated and flexible interpretation and integration into IT applications. Semantic process management can serve as an appropriate initiative to solve this problem.

The necessity of the fusion between KM and process management is a recognized issue and challenge in the literature (Records 2005). We assert that knowledge is a hidden dimension of business process, and it should be made explicit. According to the literature the ontology-based approach can be an appropriate tool for handling this issue (Karastoyanova et al. 2008; Hepp and Roman 2007). In our research and development initiatives ontology is the focal point and integration tool of BPM and KM. The following issues can be identified when investigating the knowledge-related challenges of the BPM life cycle:

- Business process strategy: changing regulations, market benchmarks, customer needs—external knowledge sources should be integrated into the process management system.
- Process documentation: collection and maintenance of knowledge elements and extension of knowledge base related to the process models should be efficient and effective, and at least semi-automated to avoid cumbersome exercises. Heterogeneous contents approaches should be integrated.
- Process analysis and design: optimization requires a detailed picture of the allocation of tasks and responsibilities, so relevant knowledge and skills should be defined; new designs will necessitate training of HR, renewed processes requires new or updated, knowledge-intensive IT applications, etc.

- Implementation and change management: knowledge content should be integrated into the new or modified systems; flexibility is necessary to integrate heterogeneous models and content.
- Process operation: handling HR changes and fluctuation, HR development requires tools that can provide customized knowledge-based training for the relevant staff. Efficient handling of deviances in the performance must be based on up-to-date and widely available knowledge.
- Process controlling/monitoring: in a complex process environment measurement can be cumbersome and result in useless reports, while a set of KPIs that are based on organizational knowledge and facilitates knowledge accumulation based on the balanced scorecard concept, customization and process intelligence can be a powerful tool for management, compliance checking and further development.

Although BPM is a complex and comprehensive approach, its scope covers strategy, organizational structure, supporting technology, skills and knowledge; it is traditionally a tool that transforms business requirements into system specifications. In the earlier waves of BPM the modelling approach of EPC (Event-driven Process Chain) was widespread methodology and organizational issues and the business view were accorded emphasis. Due to the market demand for automation in system design a more algorithmic methodology, the BPMN (Business Process Model and Notation) has become popular, and recently the overall winner of the modelling battle. Focusing on the operational logic and IT-ready descriptions of processes BPMN demonstrates an even more direct workflow orientation.

Due to its IT oriented nature the major challenge in BPM is the ability of seamless translation between the business requirements approach and IT systems and resources. Semantic Business Process Management (SBPM) is a new approach that can increase the level of automation in the translation between these two domains (Hepp et al. 2005). As process models integrate various aspects and dimensions of organizations there are serious limitations of process management in terms of maintainability, sustainability and further utilization of the typically heterogeneous content for value added services.

There is no doubt that BPM is not simply a model-based representation of organizational rules and regulations, while the utilization of models exclusively for IT development is a very limited approach. BPM can be the core of various knowledge-oriented systems through the facilitation of semantic interoperability of business process models and the maintenance and provision of reusable process knowledge for IT applications (Lin and Krogstie 2010). A major challenge in BPM is the management of the knowledge related to the processes portfolio. The distributed nature of knowledge represented in numerous information systems makes the integration issue even more challenging. Lin and Krogstie (2010) presents a framework for semantic annotation of processes to avoid the problem of the heterogeneity of distributed nature process models to facilitate the management of process knowledge.

BPM is a well-established method and technology for many companies, but the extension of modelling towards automated application generation, extended functionality and integration with other technologies (interoperability) is still a major trend in R&D. Recently, the focus of BPM activities has been on the implementation phase: process modelling is a tool that has to support (semi-) automatic IT development (Ternai and Török 2011), and the extension towards performance measurement, knowledge based applications, compliance checks (Namiri and Stojanovic 2007; Ternai et al. 2013), etc. is also a promising direction.

Semantic technologies have been integrated into BPM in recent decades to facilitate the automated utilization of models for the development of applications. Semantic description (machine processable representation) of process can bridge the gap between business logic and IT perspective (Hepp et al. 2005). Semantic annotation of the models also enhances the services built on process models. SBPM integrates BPM methodologies and tools with Semantic Web Services frameworks and ontology representation (Karastoyanova et al. 2008). Process mining can be used for the automatic discovery of process-related information (Alves de Medeiros and van der Aalst 2009). Management of the knowledge dimension of business process is a recognized problem, and many initiatives propose ontology-based semantics, and even fuzzy ontology to manage organizational knowledge (Alexopoulos and Gómez-Pérez 2012).

4   Business Process Management from a Knowledge Management Perspective

This section provides an overview of work related to our research. We compare this research work to our approach and also draw attention to its disadvantages. Sawy and Josefek investigated approaches to knowledge management within the context of business process redesign (El Sawy and Josefek 2013). They discussed the relationship between knowledge management and business processes, and analysed the role of knowledge management in the enhancement of business processes. They suggest the following three principles to enable knowledge creating capacity of a business process. Principle 1 "analyse and synthetize"; this deals with the increasing interactive analysis and synthesis capabilities around the process. Principle 2 "connect, collect and create" draws attention to the knowledge creation contribution of every stakeholder to the business processes. Principle 3 "personalize" means making the process customised for participants' needs. Three cases; Merill Lynch, Daimler Chrysler and Virgin were used to illustrate the three principles. However, even though these principles seem worth following, and in some cases are convincing, the authors don't go into detail how to operationalize them.

Rao and his colleagues followed a similar approach as we did in ProKEX research, but their research focus and the way of implementation is different (Rao et al. 2012). They used a formal organizational ontology, knowledge structure and source maps to assist business process re-engineering (BPR). They draw attention to obstacles to BPR, namely that during BPR only the business process itself is concentrated on in many cases while other important knowledge of the organization is not taken into account. Another issue is that there are no tools for identifying the cause of the inefficiencies and inconsistencies in BPR. Their approach includes the development and analysis of knowledge maps. Knowledge maps are generated from ontology and they provide a business process-related view of ontology in the form of taxonomy. The investigation of the knowledge map helps to identify inefficiencies in the business processes. The method was illustrated with a case study about a university campus in Jamaica. The main deficiency of their approach is that it is highly dependent upon the quality of the ontology. If the ontology is not instantiated properly the method may not reveal the problems related to the process.

Wu and Chen investigated the relationship between knowledge management investments and organizational performance (Wu and Chen 2014). They proposed a model defining knowledge management based performance for the relationships between three components: knowledge resources, business processes and organizational performance. They therefore targeted a similar research field as the one in the ProKEX research. They investigated six research questions related to the potential links for both knowledge assets and process capabilities, as well as business process capabilities. They dealt with the possible link between business process capabilities and organizational performance. Their model emphasized organizational learning as an important facilitator for the links between both knowledge assets, process capabilities and business process capabilities. Research was a quantitative, survey-based one, with a sample of 1000 manufacturing firms. Their findings particularly provide evidence to explain the knowledge-based view and the mediator of business process capabilities.

Schiele and his colleagues proposed a layered model for knowledge transfer and applied it to the area of business process modelling (Schiele et al. 2014). They divided the process of knowledge transfer into several stages, which they examined separately, to detect and identify errors more easily and to facilitate the prevention of misinterpretation. They distinguished four layers in knowledge transfer: code, syntactic, semantic and pragmatic layers. This knowledge transfer model is used in a business process management field, providing a more detailed description of the business process in order to facilitate knowledge transfer. This research has some overlapping issues with ProKEX research. Schiele's knowledge transfer model provides a more detailed, but guided business process description, which is crucial in ProKEX research. In addition, ProKEX was able to provide a prototype to test their knowledge transfer theory.

5 ProKEX: Beyond SoA

5.1 General Overview

The ProKEX research aimed to develop a complex application which addresses organizational knowledge management in terms of knowledge elicitation, knowledge representation and knowledge transfer. The goal of the ProKEX solution is to

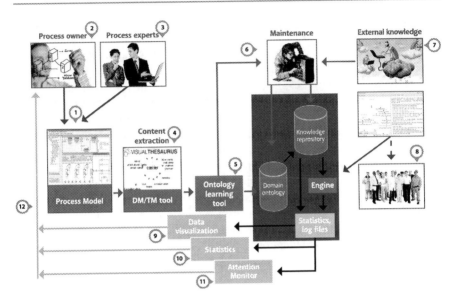

Fig. 3 ProKEX—the "Big Picture"

extract, organize and preserve knowledge embedded in organizational processes in order to (1) enrich the organizational knowledge base in a systematic and controlled way, (2) support employees to more easily acquire their job role specific knowledge, and (3) help to govern and plan human capital investment. ProKEX IT solution integrates (a) an organizational process management tool, (b) a learning management tool, (c) a monitoring and feedback tool and (d) data and text mining tools for developing a knowledge base (domain ontology) and the interfaces which are responsible for the communication between these components (Fig. 3).

The basic assumption behind the project development can be summarized as follows: organizational knowledge resides in many forms in an organization, in written form, increasingly in electronic format, such as databases and portals, as well as in a hidden form in the heads of employees. It is well known that the latter mentioned organizational knowledge is volatile, partly because for the most part it is hidden knowledge, and partly because human resources cannot be tied to the workplace forever. Organizations normally meet their strategic goals performing processes. Process by definition comprises logically grouped tasks and process or processes comply with the functions of the organization. We can start from the axiomatic definition: a task is an activity which in the value chain creates value (output) from input through using different capital assets. The logic behind grouping tasks is the business requirement, that on which level the set of output is meaningful from the point of view of the functional organization of a company or any other kind of organization. Hence we must deal with several concepts, such as value, input–output, task output and process output, interpretation of functional decomposition of value creation, strategic goals and functional strategic goals as well as process outcome. The list goes on.

Many organizations, for different purposes, try to describe their own processes. One very strong driver is the culture of quality management, to get an ISO qualification, a well-articulated process description is a must, although the static description with the need for an annual or bi-annual overview is a necessity. Far fewer companies apply the CMMI model[2] to create a sound basis for process improvement. In their case the process descriptions are also a basic precondition. An interesting and increasingly popular driver is the introduction of workflow management. Automation the processes or part of the processes assumes explicit process definitions. The most general approach of process modelling aims to generate a written, consistent regulatory document.

In any case, whatever the primary driver of process modelling is, if we wish to acquire organizational knowledge we must start on the one hand with organizational functions and regulatory documents of the performance from the point where activities can be seized in a manner which is familiar to the stakeholders. During this and other research work we found the process modelling the most appropriate segment of business and organization development to start with.

For obvious reasons, the process approach and the organizational approach are strongly connected and are therefore thoroughly mixed up in most cases. Therefore we had to set up a framework to show how the functional, process approach and organizational approach can be applied together in a coherent way. In the very first phase during the process modelling phase we focus on the human resources as performers, or the asset used to create value. Doing this provided us with a very clear criterion for the granularity of process modelling. Secondly, human resources are considered real people in different positions who are grouped in organizational units and organizations. On the other hand, their relations to the tasks to be performed are described in job roles. One person in a specific position can have one or more job roles, and vice versa, one job role can be performed by several persons.

For example, night shift is a job role, taken by a nurse, but several nurses can do the night shift, not only one. Of course, a nurse is employed not only to be on duty during the night since (s)he has many other obligations (i.e., job roles).

The first problem to be solved is: where to find the organizational knowledge, if the knowledge is strongly linked to the activity (i.e., task), but it resides—partly—in the head of the performer. To resolve this contradiction we had to introduce a very strict modelling convention. From the process and task approach, the task description is the starting point, as every task can be characterized with a list of 'to do's, the knowledge, or more generally the competencies needed to complete the 'to do', and responsibility. Responsibility is a very complex concept, for the sake of

---

[2] Capability Maturity Model Integration (CMMI) is a method to support process improvement initiatives.

simplicity we interpreted responsibility as decision making and reporting, but we are well aware that this is a major simplification.

In contrast with the usual way of process modelling, it is not the position but the job role that should be assigned to the task as performer. Following this, conventions, job roles and task descriptions can be connected to each other unambiguously. Because the job role and tasks are in 1:m relation, the job knowledge may be located either within only one task or within more tasks. Collecting and analyzing those task descriptions that belong to one job role we can conclude the relevant job knowledge. Whenever there is a change either in an assignment or in the definition of a task the requested job knowledge will change automatically. This is ProKEX's added value.

5.2   ProKEX Suite in a Nutshell

ProKEX is a complex software suite which comprises several components (see Fig. 3):

1. Process modelling (for the sake of the project, the ADONIS Community Edition is used for process modelling, but any model is suitable for the needs of ProKEX that provides standard xml output). The output (.xml) is stored in a server repository (ppmr.netpositive.hu).
2. After downloading the process models and selecting the desired one, in the next stage, the process model is analyzed through text mining solution (see details in Chapter "ProMine: a text mining solution for concept extraction and filtering"). The term 'process model' may mean only a single process or process group(s). The purpose of text mining is to extract the job-role relevant pieces of knowledge from the task description, in other words, what needs to be known for a given task in order to execute it at the expected level of quality, in time, etc. In this context the extracted pieces of knowledge will bear the task and process attributes, in terms of their identification and naming. We intentionally do not take into consideration the otherwise necessary hands-on skills, and attitude competencies, however the task/process attributes will be retained in the case of every extracted piece of knowledge. The text mining component can also be replaced by any other component that provides the same output format.
3. In the process model there is a job-role—task assignment. The more specific the job-role is, the better. ProKEX will associate every task where the same job-role occurs; hence all the knowledge associated with the task will be recalled.
4. The extracted knowledge as task specific concepts (plainly speaking a list of words) will be matched with the domain ontology.[3] The result of matching is threefold. Either (a) the matching is successful, or (b) despite the fact that the

---

[3] Detailed description of knowledge representation, namely domain ontology will be given in Chapter "STUDIO: ontology-centric knowledge-based system".

concept was not part of the list of extracted concepts the matching algorithm nevertheless discovered some relevant concepts in the domain ontology (through the ontology relations); or (c) it is unsuccessful.

(a) In the latter mentioned—(c)—case, the list of unmatched but relevant concepts will be directed to the ontology administrator, who will decide what and how to maintain and enhance in the existing domain ontology.

(b) In the case of a full-match—(a)—there is nothing to do. The matching algorithm discovered relevant concepts in the domain ontology (through the ontology relations)—and these concept are automatically put into the ConceptGroup.[4]

(c) An interesting situation comes about—(b)—when several, explicitly not mentioned, but relevant concepts are suggested for consideration. In this case the expert in charge of modelling may decide to include the suggested concept or drop it.

5. When selecting additional concepts from the suggested pool a full list will be made. The full list, as a result of the previous stage, will control the composition of the ConceptGroup. The ConceptGroup represents a specific set of the concepts related to a given job-role. The assumption here is that at least one job-role is assigned to a task, and every position is associated with at least one job-role. One job-role can be assigned to more than one position. In ProKEX, the organisational view of the process model manages the position-job-roles relation.

6. Finishing the selection, the ConceptGroup is added to the STUDIO[5] adaptive testing engine, and the STUDIO is ready for use.

7. In the final stage, the unmatched concepts can be stored in a separate text file and transferred to the ontology administrator for further elaboration.

5.3   ProKEX in Use

5.3.1   Architecture

The architecture of the ProKEX suite follows the service-oriented architecture principle. The suite consists of several components: process modelling (ADONIS), text mining (ProMine), ontology building, ontology customization, e-learning applications (STUDIO). The applications communicate to each other through web services. During the development and test phases the selected architecture provided a convenient environment, especially because the specific applications could be replaced by other applications with similar functionalities. Using standard interfaces the flexibility of the background composition will provide good opportunities for exploitation at a later date (Fig. 4).

---

[4] See Chapter "Ontology tailoring for job role knowledge" for details.

[5] STUDIO—e-learning platform developed by Corvinno Technology Transfer will be explained in detail in Chapter "STUDIO: ontology-centric knowledge-based system".

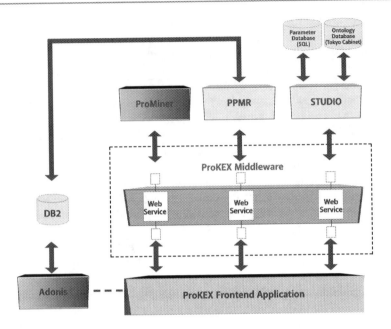

Fig. 4 ProKEX architecture

5.3.2 Usability

ProKEX is an application or to be more precise: a set of applications which is not planned for everyday use. Process management, process improvement, knowledge transfer do not occur every day. Otherwise there would be a big problem with the stability and management of the organization, and not even ProKEX could help with this. A second point: anticipated users are not naïve-users and we assume they are experts in their own areas of expertise, therefore we did not intend to produce an idiot-proof system.

ProKEX's provides several functions. In the design of the functions we tried to cover the whole life-cycle of a selected part of organizational knowledge management. The potential user or users can easily achieve an overview of the functions of ProKEX at the front-end. We have to admit that since the present status of the ProKEX is "proof of concept", the design is rather minimalist, and for commercial use the visual arrangement of application and content elements should be re-worked. There are plenty of opportunities for the visual design, and even for gamification, which is so trendy nowadays (Kamasheva et al 2015). The modes of interaction with application functions and content elements are mainly controlled by the back-end applications. It was not our intention to homogenize the GUIs of the applied back-end applications. Taking into consideration that different parts of ProKEX is intended to be used by different actors (users), the variability of and multiplicity of applications very likely will not invoke any problem. The use of standard interfaces and web services guarantee seamless work. The front-end substantially facilitates the navigation through application elements and content elements.

During the test phase and the feedback from the test users, the initial idea behind the development was approved. The priority of the process view was justified, and the organizational knowledge and its management have been shaped into a manageable, down-to-earth dimension. Organizational knowledge is a very loosely defined concept, hence managing organizational knowledge spans from the yellow-book like data inventory, through simple or advanced full-text search or document indexing facilities up to the data—and text mining application. Since the applied methods always strongly depend on the purpose of the application, no one can say this or that is worse or better. Following the "fit-for-purpose" principle, in ProKEX the stress is on eliciting, representing and transferring organizational knowledge. This is done in order to meet strategic requirements, such as increasing productivity, mitigating risk and minimizing loss from the high volatility of a skilled and knowledgeable workforce.

5.3.3   Process Modelling

There is a fairly broad range of freedom in selecting the level of granularity in process modelling. It is very difficult to tell theoretically what the optimal level is, however, we never encountered this problem in a real-life environment. There is a rule of thumb that the task still to be modelled should have at least one human performer. Another orienting principle is that the modelled processes somehow should be suitable for the functional decomposition of the organization; hence the organizational approach cannot be ignored. In the case of a bureaucratic structure the process group structure will clearly be different than in a divisional or in a matrix structure, not to mention virtual organizations. These aspects should be taken into consideration before starting detailed modelling.

5.3.4   Text Mining

The question is how can we use text mining to extract information/knowledge/concepts from processes in order to enhance or populate the existing ontology? The main problem that we address connecting text mining to process management is the following: Process modelling focuses on tasks (that is by definition an activity with attributes—minimum—I/O, resources allocated to the execution, and competencies [put simply: the need to know]). The next most important issue in process modelling is: how the tasks relate to each other, the flow, what comes first, what next, where they are executed in parallel, what triggers the execution of a specific task. Third question: what tasks belong to what process (this can be understood from the business logic, e.g., accounting, logistics, etc.). The forth main issue: how the processes are connected to each other, what comes first, what next...For example a product must be first manufactured, then dispatched and not vice versa. Again, the answer is in the business logic.

Why is text mining used? As transpires from the brief description of the nature of process modelling it is rather procedural, while many questions raised by the modelling need to be answered on a contextual basis, where the context has a rather declarative nature (in our case the STUDIO ontology). The text mining application with its simple or sophisticated procedures connects the two different approaches, processes the concepts, and transfers them to the ontology for enhancement of the same. The purpose of ontology building and enhancement is not for its own sake but to provide a contextual background, (this was referred to above as business logic) which is necessary for the process modelling (later improvement and optimization).

The text mining component addresses the following issues: (i) how to extract the knowledge from the processes, namely from the task descriptions, and (ii) how to enrich the extracted concepts in order to improve the efficiency of ontology matching and/or ontology enhancement. ProMine,[6] the text mining component extracts knowledge elements by using domain corpuses and lexical databases. In order to gather domain related concepts among text data it uses a semantic similarity measure that is a combination of statistical and semantic approaches. The text mining application provided very good results in the sense of enrichment of the concept inventory, which we initially built up from the processes. The level of enrichment may vary depending on what domain corpus was used, but of course it is very much domain dependent. In the case of insurance the supply of domain corpuses is much larger than in the case of EIT Fund Management.[7] In the latter case only the official guidelines and handbooks played a role.

The novelty of the solution may be considered as a new approach for process mining. Processes are analyzed with data and text mining techniques to extract knowledge from the tasks in order to match them to an organizational knowledge base. In this way the organizational knowledge base will conform with the process structure to some extent. In our approach, the knowledge base is an ontology, which provides the conceptualization of a certain domain. The main innovation lies in new algorithms for the extraction and integration of the static and dynamic process knowledge, through enrichment of the extracted task knowledge as will be discussed in detail in Chapter "ProMine: a text mining solution for concept extraction and filtering". In order to avoid ambiguous concept generation and overwhelming concept matching, similarity and information gain measurements are introduced and used.

5.3.5   Ontology Tailoring

Ontology tailoring can be considered as a type of machine learning. We would claim this component is at the heart of ProKEX, since the success of knowledge transfer (one of the KPI's) largely depends on which concepts can be identified in the ontology and which ones are missing. The missing elements are a clear indication for the ontology enhancement of how hidden organizational knowledge is discovered and represented. The efficacy of ontology tailoring depends on two

---

[6] ProMine will be explained in details in Chapter "ProMine: a text mining solution for concept extraction and filtering".

[7] See Sect. 5.4.2.

factors: depth and similarity. When a concept is extracted from a task, the concept is identified in the ontology, and due to the nature of the ontology few or many underlying concepts are also identified as required knowledge in the context of a given task and job role. The efficiency of ontology tailoring depends on the selected depth to drill down. If the selected level of depth is too high, the knowledge transfer (in the form of a knowledge test) will be too detailed. Conversely, if it is too small, then the knowledge test and transfer will remain on the surface, and the strategic advantage of the system will cease to apply. There is no general rule for a good decision, though we found on an empirical basis that depth $= 3$ provides satisfactory results.

One interesting feature of the ProKEX in the area of ontology tailoring is the option of adding additional knowledge elements to those which seem to be identically identified. Because of the richness of naming, whatever convention we follow it is likely that there will be more similar or less similar concepts worth including in the customized ontology; applying a measure of similarity is a major challenge. However, it was not applied in the present implementation and thus remains one of the future development targets.

5.3.6   Job Role: Task Assignment

In the BPM world, RACI (Responsible, Accountable, Consulted, and Informed) is widely used. Its origin goes back to the 20s of the last century and it is supposed to show what kind of role a performer plays in a specific job. Depending on the variety of the roles many dialects of the RACI have proliferated (PACSI, RASCI, RASI, RACIQ, RACI-VS, CAIRO, DACI, RAPID, RATSI, etc.). In ProKEX we use the terms with a different meaning as explained earlier. A job can be precisely defined by a task description (procedural description, what to do?), competencies (what to know?) and responsibility, autonomy (how to make decisions, whom to report to?). A job role needs to be distinguished from the position, which in our interpretation is membership within the organization and always belongs to a person. Hence one employee may be suitable for one or more job role(s), and conversely, the same job role may belong to one or more position(s).

5.3.7   Evaluation in Terms of Statistics

The STUDIO knowledge test provides not only customized learning material but also a plethora of statistical results. Even if the test is not an "against-the-clock" type, the time consumed for the test, and break down according to by questions/ nodes already reveals something, especially if there are several similar test candidates, and the mean of the time consumed and deviation inform us about the difficulty or potential irregularities. More emphasis can be applied to the properly answered questions/total questions indicator as an overall performance indicator. In addition to this the properly answered questions/total questions related to the underlying concepts indicators are also available. It is not only the performance per test candidate that is important since the distribution of incorrect answers is also meaningful. It highlights systematic errors, and also gives some idea where and which in-house or other training is necessary.

5.3.8   Process Improvement

In the life-cycle approach one kind of potential feedback may be used to process improvement. As mentioned earlier, some additional concepts were already identified through the ontology matching, and these additional knowledge elements can be added to the task description. This is the process owner's responsibility. Adding additional knowledge elements to the job role is the responsibility of the human resource management (HRM). For the time being we cannot fully automate the additions, however, under the control of the process's owner the additional task description elements can be added to the process model as an information object.

5.3.9   Ontology Enhancement

Ontology enhancement is a manual activity in the long term since we manage one consistent ontology, while maintaining the consistency of the ontology is of primary interest. In order to carry out the enhancement, input is derived from the text mining: knowledge elements are dug up from the task descriptions and filtered via the ontology matching as unmatched knowledge elements.

5.4   Use Cases

In ProKEX we investigated three use cases. The reason for selecting different use cases was to properly test the solution to be developed. In the following subparagraphs the specifics of the selected processes are highlighted.

5.4.1   Food Chain Safety Sampling Process

The globalisation of the food industry increases the complexity of the food chain and establishes new requirements for the food safety. As the food supply chain crosses many borders, following the 'from farm to fork' principle where regulations and inspections cover each phase of production it is a complicated and knowledge-intensive task. New challenges arising for the stakeholders in food safety mean that the Food Authorities must adopt new approaches. A critical factor is to ensure that the officers involved in food standards enforcement have the right kind of knowledge and competencies.

In collaboration with the National Food Chain Safety Office (NEBIH), the sampling process was selected as a testbed for the pilot of ProKEX. NEBIH is the Food Authority in Hungary responsible for food safety along the entire food supply chain. NEBIH has its own sampling policy and an annual sampling programme. Thousands of on-site sampling and inspections are carried out each year by the geographically dispersed task forces of NEBIH.

The basic requirement of any regulations concerning sampling is that they include details of the steps to be taken. Detailing the stages of the process is no trivial task due to the complexity of the food chain and the wide range of products and substances that are sampled regularly. The general phases of sampling can be classified according to the order of the activity: preparation—before going out sampling, on-site sampling, procedures after sampling, follow-up—dealing with the results of the sampling. There are various policies and regulations for various products and substances, and these can vary depending on the phases of sampling.

The complexity of sampling in terms of extensive regulations and the geographically dispersed organisational structure of NEBIH demand a new approach that would focus on knowledge management. It is vital to maintain the same level of knowledge for every person involved in sampling, and, at the same time, to transfer any up-to-date knowledge to them. The new approach ensures not only knowledge transfer but also promotes common understanding.

5.4.2   Fund Management

The European Institute of Innovation and Technology (EIT) is a body of the European Union based in Budapest, Hungary. The EIT is the first EU initiative to fully integrate all three sides of the Knowledge Triangle (higher education, research and business) through Knowledge and Innovation Communities (KICs). The EIT's main responsibility is to promote the collaboration in the knowledge triangle by launching KICs in different domains (ICT, Climate research, Renewable Energy, Food Sciences, and Health). KICs are composed of leading universities, research labs and companies that form dynamic cross-border partnerships. Together, they develop innovative products and services, start new companies, and train a new generation of entrepreneurs.

The EIT funding model seeks to leverage and align innovation investment. Therefore, on average, the EIT financial contribution does not exceed 25 % of a KIC's overall funding. The EIT financial contribution to the KIC is provided primarily in the form of a grant for action, covering activities contributing to the integration of the Knowledge Triangle of research, innovation and higher education described in detail in a Business Plan.

The Grant Management Cycle

The process for the "Allocation of funding" lasts about 1 year (in terms of lead-time) and involves different actors: EIT Officers, Governing Board Members, KICs and experts. The macro phases of such a process are:

- Definition of policies and guidelines. In the first phase the EIT guarantees the rules that will govern the competition among the KICs. EIT shall draft the guidelines for the preparation of the business plan and the rules for the allocation of funding exercise.
- Analysis of KICs' past performance. To produce an assessment of the KICs' past performance the EIT-HQ analyze the outputs of the assessment of the reporting for previous years and give an evaluation based on the rules defined in the previous phase.
- Analysis of KIC annual business plan. Experts are contracted to evaluate the business plan presented by the KICs according to the three pillars (education, entrepreneurship and research) and the merit of the thematic area of each KIC.

Experts then provide a forecast for the quality of such proposals according to the rules defined by the EIT Governing Board.

- Hearings of the KIC in a multi annual perspective. This process is very critical because it is the formal process that allocates most of the budget of the EIT. Compared to other similar programs the allocation is made on the basis of competitiveness.[8]

Process of the Evaluation of the Business Plan for KICs

In this phase, experts are selected and contracted to evaluate the activities that the KIC decides to carry out in the following year. The evaluators assess the business plan produced by the KICs according to the modalities that the EIT Governing Board has defined in the first phase by carrying out an evaluation according to the various domains. Although the process is quite stable, the EIT Head Quarter has to adapt it according to the decision of the Governing Board for the specific year.[9] This is one of the areas where the adoption of the ProKEX platform will provide a specific benefit in order to decide if the available knowledge is sufficient to perform an appropriate and independent assessment.

5.4.3 Insurance

Our project partner is a middle-sized, Hungarian insurance company operating both in the Life and Non-Life line of insurance business. The insurance company is relatively young; it was founded by Hungarian stakeholders only 8 years ago. In the course of the Insurance pilot of the ProKEX project, we have modelled close to 100 processes of an insurance company. For the case study we selected two complex processes that help us envision the proposed solution.

Loss Claim Management

The first process is the loss claim management of the Non-Life branch. Loss claims arise either from new claims by the insured parties or from reactivating a claim when new information emerges regarding the issue. Every aspect of the issue is collected in a virtual claim issue file. The process starts with the inspection of the incident. The first and foremost information to collect relates to whether personal injuries are involved or not. In most cases, especially if the estimated loss exceeds a given limit, an inspection or a verification of evidences is necessary. This is undertaken by subcontracted inspectors, who are the experts in the issue's insurance coverage field. If the amount of loss is determined a decision mechanism within the insurer organization is triggered that results in a final decision on the claim. When the insurer decides the magnitude and other conditions of the disbursement, an administrative sub process takes place involving the notification of the stakeholders of the issue, the decision on the further existence of the insurance contract, and the handling of the effective payment of the disbursement. If the loss results in the

---

[8] The EIT Governing Board defines the percentage of competitive funding yearly.

[9] The business case took the 2014 Allocation of funding implemented in 2013 into consideration.

contract becoming obsolete, (e.g. a vehicle is deemed a total loss), the insurance contract is discontinued by the insurer, which might require further action in settling overdue or overpaid balances. If a third party is involved, and the loss claim has been fully undertaken by the insurer, a regression process starts that attempts to identify the insurer of the third party and to negotiate based on the legal regulation or bilateral agreements between the insurers.

New Insurance Offer

This starts by creating a personalized offer for a prospective insured party and ends with the contract being signed or rejected. A request for a new offer always originates from an agent or a representative distributor of the insurance company. The first task is to determine the identity and the eligibility of the main parties of the offer, the beneficiary parties, and the agent eligible for commission. All available information about the parties is recorded on the issue with special care being made to maintain data integrity, duplication elimination and data quality management. The agent involved in the offer, being the active insurance provisioning partner of the insurer, has to be contracted. The examination of the agent includes a thorough inspection involving a designated scoring method, including the calculation and update of the so-called "ABC indicator", which qualifies the agent based on the commission balance, the outstanding premiums of the agent's contracts, and the rate of early contract deletion. The offer issue continues in two parallel threads: health and financial risk assessment. Based on the conditions of the offer and the regulations of the insurer, the administrator has to decide whether it is necessary to conduct a health risk evaluation. In this case the issue is handed over to the designated health risk assessment team. The health risk evaluation can take place simply based on the available documentation and statistical data, or it might require a medical examination of the life insured parties. If the medical examination is necessary it must be ordered from a third party service provider. At the end of the sub process the team submits a recommendation to the new business administration where a decision is made that in some cases includes the insurer's lead medical expert.

The term for the financial risk assessment in the insurance domain is prevention. The aim of the prevention sub process is twofold: it stops the customer from undertaking a financial commitment that is beyond his/her financial means, and also protects the insurer from entering into a contract that is likely to fail abortively. If both types of risk assessment have been successfully concluded the new business department makes sure that all the necessary proclamations and statements have been received by the insurer.

5.5 Future Development

The ProKEX solution has more potential for future development (Gábor et al. 2013). The process model export enables the creation of process ontology. Process ontology (see Chapter "Corporate semantic business process management" for

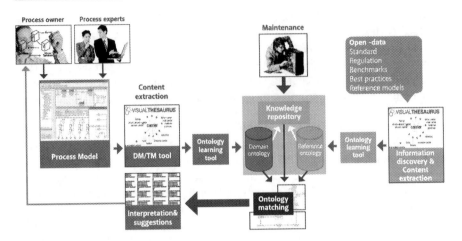

Fig. 5 ProKEX future development opportunities

details) differs from the domain ontology, since it is rather the ontological representation of the processes including tasks and task attributes than the representation of the underlying concept hierarchy. Process ontology may serve comparison purposes very well. An actual process can be compared with any other process claiming a similar functionality or process goal. The objective of the comparison can be either process improvement or checking for compliance (Fig. 5).

During the process improvement the deviation of the process in question from those ideally taken from the inventory of best or good practices will show the process owner in which direction to initiate changes. Even if the process owner receives a highly detailed gap analysis, the changes cannot be automated since there are several other preconditions which are not presented in the process ontology, such as organizational, cultural, regulatory, etc. to be taken into consideration.

If the actual process is compared with standards (e.g. ISO), or strongly recommended guidelines (e.g. COBIT), the comparison can lead to a more strict compliance checking, and in this way the solution can very efficiently support any kind of accreditation or certification activity. Due to the fact that numerous standard processes are a compulsory part of corporate processes (taxation, information security, quality assurance, to mention but a few), the ontological support of compliance checking cannot be overemphasized.

An interesting question is from where the reference process should be taken. Considering the development of data management, as well as the growing number of open data applications and sources, it is expected that the reference process will be increasingly accessible in a standard format, and available to create reference process ontology. From this point the matching of the two ontologies will be a matter of using or developing the appropriate techniques, which is still a big challenge.

References

Alexopoulos, P., & Gómez-Pérez, J. M. (2012, May 27–31). Dealing with vagueness in semantic business process management through fuzzy ontologies. In *Proceedings of the 7th International Workshop on Semantic Business Process Management, Heraclion, Greece.* http://sbpm2012.fzi.de/images/SBPMp4.pdf

Alves de Medeiros, A. K., & van der Aalst, W. M. P. (2009). Process mining towards semantics. In T. S. Dillon, E. Chang, R. Meersman, & K. Sycara (Eds.), *Advances in web semantics I* (Lecture notes in computer science, Vol. 4891, pp. 35–80). Berlin: Springer. http://dx.doi.org/10.1007/978-3-540-89784-2_3.

Anonymous. (2005, June). *ARIS value engineering-concept. Whitepaper.* IDS Scheer AG. http://www.sdn.sap.com/irj/scn/go/portal/prtroot/docs/library/uuid/eae8e311-0b01-0010-0f9c-8d26e2714a91?QuickLink=index&overridelayout=true&5003637725232

Argyris, C. (1992). *On organisational learning.* Oxford: Blackwell.

Argyris, C., & Schön, D. (1978). *Organisation learning: A theory of action perspective.* Reading, MA: Addison-Wesley.

Child, J., Faulkner, D., & Tallman, S. (2005). *Cooperative strategy—Managing alliances* (2nd ed.). Oxford: Oxford University Press.

Child, J., Markóczy, L., & Cheung, T. (1994). Managerial adaptation in Chinese and Hungarian strategic alliances with culturally distinct foreign partners'. *Advances in Chinese Industrial Studies, 4,* 211–231.

Cohen, W. M., & Levinthal, D. A. (1990). Absorptive capacity: A new perspective on learning and innovation. *Administrative Science Quarterly, 35,* 128–152.

Davies, J. F., Grobelnik, M., & Mladenic, D. (Eds.). (2009). *Semantic knowledge management. Integrating ontology management, knowledge discovery, and human language technologies.* Berlin: Springer.

Dixon, N. (2000). *Common knowledge.* Boston: Harvard Business School Press.

Easterby-Smith, M. (1997). Disciplines of organizational learning: Contributions and critiques. *Human Relations, 50*(9), 1085–1113.

Easterby-Smith, M., & Lyles, M. A. (2011). *Handbook of organizational learning and knowledge management* (2nd ed.). Chichester: Wiley.

El Sawy, O. A., & Josefek, R. A., Jr. (2013). Business process as nexus of knowledge. In C. Holsapple (Ed.), *Handbook on knowledge management 1: Knowledge matters* (Vol. 1, pp. 425–436). Berlin: Springer.

Fiol, C. M., & Lyles, M. A. (1985). Organizational learning. *Academy of Management Review, 10*(4), 803–813.

Gábor, A., Kő, A., Szabó, I., Ternai, K., & Varga, K. (2013). Compliance check in semantic business process management. In Y. T. Demey & H. Panetto (Eds.), *Lecture notes in computer science* (Vol. 8186, pp. 353–362). Berlin: Springer.

Gábor, A., & Szabó, Z. (2013). Semantic technologies in business process management. In M. Fathi (szerk.), *Integration of practice-oriented knowledge technology: Trends and prospectives* (pp. 17–28, 368 p.) Berlin: Springer. ISBN: 978-3-642-34470-1.

Gavin, D. A., Edmondson, A. C., & Gino, F. (2008). Is yours a learning organization. *Harvard Business Review, March,* 109–116.

Gold, A. H., & Arvind Malhotra, A. H. S. (2001). Knowledge management: An organizational capabilities perspective. *Journal of Management Information Systems, 18*(1), 185–214.

Hedlund, G. (1994). A model of knowledge management and the N-form corporation. *Strategic Management Journal, 15,* 73–90. Special issue: Strategy: Search for new paradigms.

Henderson, J. C., & Venkatraman, N. (1993). Strategic alignment: Leveraging information technology for transforming organisations. *IBM Systems Journal, 32*(1), 4–16.

Hepp, M., Leymann, F., Domingue, J., Wahler, A., & Fensel, D. (2005, October 12–18). *Semantic business process management: A vision towards using semantic Web services for business process management.* ICEBE 2005. IEEE International Conference on e-Business Engineering, pp. 535, 540. doi: 10.1109/ICEBE.2005.110.

Hepp, M., & Roman, D. (2007). An ontology framework for semantic business process management. In *Wirtschaftsinformatik Proceedings 2007* (Paper 27).

Hsu, I., & Sabherwal, R. (2012). Relationship between intellectual capital and knowledge management: An empirical investigation. *Decision Sciences, 43*(3), 489–524.

Jain, A. K., & Moreno, A. (2015). Organizational learning, knowledge management practices and firm's performance. *The Learning Organization, 22*(1), 14–39.

Kamasheva, A. V., Valeev, E. R., Yagudin, R. K., & Maksimova, K. R. (2015). Usage of gamification theory for increase motivation of employees. *Mediterranean Journal of Social Sciences, 6*(1S3), 77.

Kaplan, R. S., & Norton, D. P. (1992). The balanced scorecard measures that drive performance. *Harvard Business Review, January–February*, 71–79.

Kaplan, R. S., & Norton, D. P. (1996). *The balanced scorecard: Translating strategy into action.* Boston: Harvard Business Press.

Kaplan, R. S., & Norton, D. P. (2004). Measuring the strategic readiness of intangible assets. *Harvard Business Review, 82*(2), 52–63.

Karastoyanova, D., Lessen, T., Leymann, F., Ma, Z., Nitzsche, J., Wetzstein, B., Bhiri, S., Hauswirth, M., & Zaremba, M. (2008). *A reference architecture for semantic business process management systems.* Multi konferenz Wirtschaftsinformatik, GITO-Verlag, Berlin.

Lee, H., & Choi, B. (2003). Knowledge management enablers, processes, and organizational performance: An integrative view and empirical examination. *Journal of Management Information Systems, 20*(1), 179–228.

Lin, Y., & Krogstie, J. (2010). Semantic annotation of process models for facilitating process knowledge management. *International Journal of Information System Modeling and Design (IJISMD), 1*(3), 45–67. doi:10.4018/jismd.2010070103.

Maddern, H., Smart, P. A., Maull, R. S., & Childe, S. (2014). End-to-end process management: Implications for theory and practice. *Production Planning and Control: The Management of Operations, 25*(16), 1303–1321. doi:10.1080/09537287.2013.832821. http://dx.doi.org/10.1080/09537287.2013.832821.

March, J. G. (1991). Exploration and exploitation in organizational learning. *Organization Science, 2*(1), 71–87.

Marr, B., Schiuma, G., & Neely, A. (2004). Intellectual capital—Defining key performance indicators for organizational knowledge assets. *Business Process Management Journal, 10*(5), 551–569.

Namiri, K., & Stojanovic, N. (2007). *A formal approach for internal controls compliance in business processes.* 8th Workshop on Business Process Modeling, Development, and Support (BPMDS07), Trondheim, Norway.

Nonaka, I., & Takeuchi, H. (1995). *The knowledge-creating company: How Japanese companies create the dynamics of innovation.* New York: Oxford university press.

Örtenblad, A. (2015). Towards increased relevance: Context-adapted models of the learning organization. *The Learning Organization, 22*(3), 163–181.

Pawlowsky, P. (1992). Betriebliche Qualifikationsstrategien und Organisationales Lernen'. In W. H. Staehle & P. Conrad (Eds.), *Managementforschung 2* (pp. 177–237). Berlin: De Gruyter.

Pemberton, J. D., & Stonehouse, G. H. (2000). Organisational learning and knowledge assets—An essential partnership. *The Learning Organization, 7*(4), 184–194.

Petrash, G. (1996). Dow's journey to a knowledge value management culture. *European Management Journal, 14*(4), 365–373.

Rao, L., Mansingh, G., & Osei-Bryson, K. M. (2012). Building ontology based knowledge maps to assist business process re-engineering. *Decision Support Systems, 52*(3), 577–589.

Records, L. R. (2005, September). The fusion of process and knowledge management. BPTrends. Accessed July, 2015, from http://www.bptrends.com/publicationfiles/09-05%20WP%20Fusion%20Process%20KM%20-%20Records.pdf

Scheer, A.-W., Abolhassan, F., Jost, W., & Kirchmer, M. (2002). *Business process excellence—ARIS in practice*. Berlin: Springer. http://dx.doi.org/10.1007/978-3-540-24705-0.

Schiele, F., Laux, F., & Connolly, T. M. (2014). Applying a layered model for knowledge transfer to Business Process Modelling (BPM). *International Journal on Advances in Intelligent Systems, 7*(1 and 2), 156–166.

Senge, P. (1992). *The fifth discipline: The art and practice of the learning organisation*. London: Century Business.

Skyrme, D. J. (1999). *From measurement myopia to knowledge leadership*. Access Conference, London.

Stewart, T. A. (1997). *Intellectual capital: The new wealth of organizations*. New York: Doubleday/Currency.

Sveiby, K. E. (1989). *The invisible balance sheet*. Stockholm: Affarfgarblen.

Sydler, R., Haefliger, S., & Pruksa, R. (2014). Measuring intellectual capital with financial figures: Can we predict firm profitability? *European Management Journal, 32*(2), 244–259.

Tanriverdi, H. (2005). Information technology relatedness, knowledge management capability, and performance of multibusiness firms. *MIS Quarterly, 29*(2), 311–334.

Ternai, K., Szabó, I., & Varga, K. (2013). Ontology-based compliance checking on higher education processes. In A. Kő et al. (Eds.), *EGOVIS/EDEM 2013* (LNCS, Vol. 8061, pp. 58–71). Heidelberg: Springer.

Ternai, K., & Török, M. (2011, September 8–11). *Semantic modeling for automated workflow software generation—An open model*. 5th International Conference on Software, Knowledge Information, Industrial Management and Applications (SKIMA 2011), Benevento, Italy.

Tobin, J. (1969). A general equilibrium approach to monetary theory. *Journal of Money, Credit and Banking, 1*(1), 15–29.

Volberda, H. W. (1996). Toward the flexible form: How to remain vital in hypercompetitive environments. *Organization science, 7*(4), 359–374.

Warren, P., Davies, J., & Simperl, E. (Eds.). (2011). *Context and semantics for knowledge management. Technologies for personal productivity*. Berlin: Springer.

Wu, I., & Chen, J. (2014). Knowledge management driven firm performance: The roles of business process capabilities and organizational learning. *Journal of Knowledge Management, 18*(6), 1141–1164.

Corporate Semantic Business Process Management

Katalin Ternai, Mátyás Török, and Krisztián Varga

1 Overview of Semantic Business Process Management

1.1 Business Process Management

Modern organizational trends are modulating the focus of many businesses to reorganize themselves around their business. The trends in the new networked economy make business processes and the management of these processes more dynamic and knowledge intensive than in (Weske et al. 2004). The Gartner Group predicted that by 2015 (Light 2005) there would be an explosion of interest in business process management suites and their integration with underlying software infrastructure.

In the dynamic business environments, complex organizations emphasize the importance of Business Process Management (BPM). By managing processes with continuous improvements, while the organization can reduce costs, increase efficiency, and strengthen the ability to respond to change (Weske et al. 2004). Many companies already use BPM efficiently to increase their operating flexibility. Managing business processes means focusing on the important activities and resources of a company, such as: markets, strategy, people, financial aspects, material management, intellectual properties, data and information. The aim is to

K. Ternai (✉)
Corvinno Technology Transfer Center Ltd, Budapest, Hungary
e-mail: kternai@corvinno.com

M. Török
Netpositive Ltd, Budapest, Hungary
e-mail: torok.matyas@netpositive.hu

K. Varga
Corvinus University of Budapest, Budapest, Hungary
e-mail: kvarga@informatika.uni-corvinus.hu

© Springer International Publishing Switzerland 2016
A. Gábor, A. Kő (eds.), *Corporate Knowledge Discovery and Organizational Learning*, Knowledge Management and Organizational Learning 2,
DOI 10.1007/978-3-319-28917-5_2

design and control the organizational structures in a very flexible way so they can rapidly adapt to changing conditions.

Business processes are often modeled using informal graphical methods because these techniques are perceived to be more intuitive to common users. BPM systems facilitate the management of business processes using graphical process models (van der Aalst et al. 2000). These models are unique because they are derived from graph theory formalisms. They use mathematical modelling that controls business processes in ways that other enterprise systems cannot (Basu and Blanning 2000; Curtis et al. 1992). Informal graphical modelling techniques, such as flowcharts do not permit mathematical analysis and control.

Several formal graphical process model techniques have been developed for BPM Systems e.g. Petri nets, state charts, Unified Modelling Language (UML) diagrams, Business Process modelling Notation (BPMN) and Business Process Execution Language (BPEL) diagrams.

A formal graphical process model should not only be comprehensive but must also be easy to understand because manual organizational activities are involved and so that it can be used as a platform for communication between various business people (Curtis et al. 1992).

BPM standards and specifications are based on grounded BPM theory and are eventually adopted into software and systems (van der Aalst et al. 2000; Basu and Blanning 2000).

Business process modelling has a very large literature; nevertheless there are different views, concepts and misconceptions in this area. The various Business Process Management solutions provide different modelling approaches, but the basic logic behind the modelling methods remains the same. The various approaches include the definition of activities, descriptions, and responsible positions or roles for execution. While process modelling is a traditional and well-grounded topic, the various possible motivations for modelling a process, the various sources of models, and the resulting variety of requirements on the formalisms used for representing processes are often not considered.

BPM applications are used to describe the organizational processes, together with the required information and other resources (including human resources) needed to perform each activity. Business processes are defined as sequence of activities. Each elementary task should have an organizational actor to perform it. A well described process model contains all the relevant tasks and their description. In our opinion it is necessary to unambiguously define who is responsible for the execution of each activity in terms of the RACI (Responsible, Accountable, Consulted, Informed) matrix (Jacka and Keller 2009), bridging the organizational model and the process model. Generally BPM methodologies' requirements are satisfied with the definition of the type of job role, emphasized in the RACI matrix. In our approach the job role is interpreted as a bridge between the task and the actor. One or more job roles are assigned to a position, the positions fill up the organization. The position and job role may relate to each other in several ways (1:1, 1:m, n:1, m:n) (Gábor and Szabó 2013).

One of the objectives of BPM is the transformation of informal knowledge into formal knowledge and facilitates its externalization and sharing (Kalpic and Bernus 2006). The relevant knowledge is embedded and strongly related to the roles as a building element of the organizational structure. The competences relates to the job role, considered as content. Competences mean knowledge, skill and attitude that are necessary for sufficient execution. The knowledge extraction refers to the content, while the type of the job role has more organizational aspects than knowledge management. In order to properly include the job role knowledge into the process model an extended RACI matrix should be used, where the description of task from the knowledge perspective is added to the RACI. In a turbulent environment both the roles and required competencies are changing, therefore the knowledge articulation cannot be independent from the permanently updated business process model.

BPM stages include modelling and analyzing the current process as well as the optimizing and redesigning of new processes. Process design is, therefore, a continuous process for several reasons, for example:

- New organizational concepts can arise.
- New Best Practice cases become available as reference models.
- New technologies are invented.
- New knowledge is obtained from processes, which have just been implemented, leading to an adjustment of the process.

BPM includes process engineering (design and modelling), execution, monitoring, optimizing and re-engineering. An additional feature of these applications of process modelling is the ability to simulate.

BPM Systems designed to allow the direct control of the business processes by operational level managers. This unique feature has allowed managers to monitor, change, and rapidly adapt business processes and data flows to meet the changing needs of dynamic business environments despite these managers being geographically dispersed (Weske et al. 2004; Light 2005; Basu and Blanning 2000).

It is not easy to analyze business processes, or to define and install them because a lot of business information, such as information about events, actors, conditions and artifacts are needed to understand the process. If businesses and business strategies are changing, the underlying business processes also have to be changed and adopted. Once a model of a business process is available, various analytical methods can be used to check if the process delivers the product or service in the most optimal and cost-effective way. In particular, each task can be analyzed to ensure its added value to the business and to prevent the waste of time and resources (Weske et al. 2004).

BPM is also an approach for managing the execution of IT supported business operations using the managerial process approach. In general, BPM Systems use formal graphical process models for three levels of abstraction:

- the business level,
- the execution level,
- the evaluation level.

The business level graphs that define business processes can be transformed into execution graphs. The executions of business processes can be evaluated, and by using the results the business graphs can be improved (Basu and Blanning 2000).

Formal graphical process models based on a meta model can be used as a starting point for the development of workflow-based applications. These process models must be comprehensive, understandable and formal at the same time (Green and Rosemann 2000).

BPM Systems support the collection and integration of real-time information by interfacing with a variety of enterprise systems, architectures, and technologies (Harmon and Hall 2006; Vernadat 2002).

BPM Systems integrates several major IT components and areas of research, including (Harmon and Hall 2006):

- process modelling tools,
- simulation tools,
- business rule management tools,
- BPM applications,
- business process monitoring tools,
- software modelling and development tools,
- enterprise architecture integration tools,
- workflow management tools,
- business process modelling languages,
- organization and enterprise modelling tools.

1.2   Semantic Business Process Management

In spite of BPM having attracted significant attention from both research and industry, the degree of mechanization in BPM is still very limited and does not provide a uniform representation of an organization's processes on a semantic level, which would be accessible to semantic functions, such as intelligent queries (Lautenbacher and Bauer 2006). In this respect BPM tools and techniques include fundamental problems such as:

- difficulty in querying and reusing business processes (Hepp et al. 2005),
- inability to automatically transform a business process model into an executable workflow model (Basu and Blanning 2000),
- lack of semantic description in business process execution language specifications, such as BPEL for dynamic discovery and automatic composition of web services (Hepp et al. 2005),
- difficulty in integrating business processes across organizations (Hepp and Roman 2007; Hoefferer 2007),

- difficulty in the connection between static and dynamic process data (Gábor and Szabó 2013).

Semantic web technologies and semantic web services technology provide suitable large-scale, standardized knowledge representation techniques to over-come the above mentioned problems. The term semantics means the study of meaning in language, or the study of relationships between signs and symbols and what they represent. It also indicates the meaning or the interpretation of a word, sentence, or other language form (Fensel et al. 2005). Fensel and his colleagues have proposed combining the Semantic Web field, the BPM and the provided consolidated technology, which they have dubbed semantic business process management (SBPM) (Fensel et al. 2005; Hepp et al. 2005).

SBPM is a new approach increasing the level of automation of BPM, for example, in the translation between business and IT. The basic idea of SBPM is to combine Semantic Web Services frameworks, ontology representation, and BPM methodologies and tools, and to develop a consolidated technology (Karastoyanova et al. 2008).

Ontology definition is the key element in providing a visual and textual repre-sentation of the processes, data, information, resources, collaborations and other measurements. Several authors have drawn parallels between the ontologies and the role of XML in data representation. Ontology is responsible for conceptualization and for structuring knowledge embedded in business processes. Ontologies are state-of-the-art constructs to represent rich and complex knowledge about things, their properties, groups of things, and relations between things.

The use of web-based ontologies and their contribution to business innovation has received a lot of attention in recent years (Berners-Lee et al. 2001). Ontologies provide the means to freely describe different aspects of a business domain, and basically provide the semantics making it possible to describe both the semantics of the modelling language constructs as well as the semantics of model instances. It describes not only data, but also the regularity of connection among data.

The most important description language of the semantic web is the OWL (web ontology language) preferred by W3C (Hepp et al. 2007). With web-based semantic schema such as the OWL, the creation and the use of specific models can be improved, furthermore the implicit semantics contained in the models can be partly articulated and used for processing. The goal is to be able to apply machine reasoning for the translation between the spheres, in particular for the discovery of processes, process fragments and for process composition (Benjamins et al. 1996).

The use of ontologies is a key concept that distinguishes SBPM from conven-tional BPM. The role of ontologies in SBPM means emphasizing the opportunity to embed process structure information in ontologies. Ontologies are used to structure its underlying knowledge and enable comprehensive and transportable machine understanding. They facilitate knowledge sharing and reuse between various agents, regardless of whether they are human or artificial (Fensel et al. 2005).

The principle of ontological completeness states that there needs to be a direct relationship between the design constructs used in graphical process models and the ontological real world constructs they represent (Wand and Weber 1993). From the ontological completeness perspective, design constructs are symbols, notations, while semantics (graphical process model constructs) are used to explicitly map ontological (real world) constructs. These design constructs can be interpreted according to the meanings of the ontological constructs in the real world from the users' individual aspects. The completeness means that a graphical grammar used by graphical process models must contain constructs that enable it to model any real world entity in which a user is concerned. When reading a symbol users should be able to comprehend the information stored in it. With this object an unequivocal relationship between the graphical symbol and its meaning in the real world has to exist (Wand and Weber 1993).

The object of SBPM is to support the flexible and efficient implementation of BPM by bringing semantics to the business processes so that both the business and IT worlds can traverse them without too much physical effort (Hepp and Roman 2007). A number of studies related to SBPM have attempted to carry out the aim of SBPM in an effort to realize the initial promise of BPM. Hepp and Roman (2007) proposed upper level ontologies associated with business processes (e.g., organization and resources, business functions, logics and strategy) by listing some informal competency questions (Hepp and Roman 2007).

Some research in SBPM has primarily dealt with the representation of a semantically annotated business process model by incorporating semantics into specific business process models created using specific modelling methodologies (Scheer et al. 2005). The objectives of these studies are business process integration (Lautenbacher and Bauer 2006) and the semantic extension of EPC modelling methodology (Thomas and Fellmann 2007). The focus in these studies is on the semantic and representational differences in the design of business processes in different organizations, which means different terms, different modelling notations, and different representations of the same business process. Building semantically rich business processes may appear to be a costly, time-consuming, and complex task. However the resulting knowledge in processes, once created and continually managed, can be highly useful in both the business world and the IT world. Thereby flexible and efficient BPM can be achieved by reducing the time and cost involved in developing new business processes. The semantically richer business process information makes it possible to check stronger conditions.

Some other research has focused on examining business processes using semantic technology such as ontology. Celino and his colleagues introduced several technologies for semantic business process analysis, including process mining and reverse business engineering, and described how those technologies could benefit from the use of semantic information (Celino et al. 2007). Pedrinaci and Domingue developed event ontology to support the monitoring of events at a specific time and process mining ontology to integrate diverse knowledge that can be utilized to mine business processes (Pedrinaci and Domingue 2007).

The generation, processing and visualization of ontologies are supported by an extensive set of tools and frameworks. This general but formalized representation can also be used for describing the concepts of a business process.

Within SBPM two types of ontologies are utilized: domain ontologies and process ontologies. Domain ontologies support process modelling, amongst others, in terms of describing the actual data that is processed during process execution. Through this semantic description of the data business process analysis can be semantically enhanced since the semantic meaning of the data is preserved during every phase of the process lifecycle (Herborn and Wimmer 2006).

According to our current knowledge, process ontologies have no precise definition in academic literature. Some refer to process ontology as a conceptual description framework of processes (Koschmider and Oberweis 2005). In this interpretation process ontologies are abstract and general. In contrast to this, task ontologies determine a smaller subset of the process space, and the sequence of activities in a given process (Gábor and Szabó 2013).

The domain ontology provides vocabulary of concepts and their relationships, and captures the activities performed on the theories and elementary principles governing that domain. Process ontology identifies all the artifacts that describe a process, regardless of whether it is structured or not. It makes it possible to clearly and unambiguously build all the process elements linked with the domain ontologies that specify enterprise concepts, as well as the business rules, roles, outcomes, and every other interdependency.

In our approach the concept of process ontologies is used, where ontology holds the structural information of processes with multi-dimensional meta-information partly to ground the channeling of knowledge embedded in domain ontologies. The attempt is to undertake the tasks and provide an extension for the standard ontology definition in the form of an annotation scheme to enable ontologies to cover all the major aspects of business process definition.

The chapter focuses on the SBPM aspects of the solution utilized in the ProKEX[1] project. We demonstrate a semi-automatic methodology to extract, organize and preserve knowledge embedded in business processes to enrich organizational knowledge base. In the semantic approach, the only thing we can handle operationally is the piece of knowledge which is necessary to complete the given process stage. The solution is based on the connection between the process model and corporate knowledge base, where the process structure will be used for building up the knowledge structure in an ontology. We discuss how to establish the links between model elements and ontology concepts. The objective of this approach is to transform the business process into process ontology and to combine it with the knowledge base as a domain ontology in a dynamic, systematic and well-controlled solution.

---

[1] ProKEx: Integrated Platform for Process-based Knowledge Extraction, EUREKA project, http://prokex.netpositive.hu.

In the next sections, the examples will be outlined as a proof of evidence. In the case study we illustrate the solution related to the processes of a medium-sized, Hungarian insurance company operating both in the Life and Non-Life line of insurance business.

2 Knowledge Extraction from Process Models

The current chapter describes the proposed solution for capturing every aspect of a business process, extended with the identification and mapping of the knowledge items. The modelling procedure set forth in this section is applied in the case study.

2.1 Business Process Modelling

Business Process Modelling is the first phase of the Business Process Management lifecycle. In the ProKEX project business process models were implemented by using the BOC ADONIS modelling platform (BOC Group 2013). We selected this tool because of its popularity in modelling practice. However, our approach is transferable to other semi-formal modelling languages such as ARIS, etc.

ADONIS is a graphical Business Process Modelling language. The main modelling object is the activity. The ADONIS modelling platform is a business meta-modelling tool with components such as modelling, analysis, simulation, evaluation, process costing, documentation, staff management, and import–export. Its main feature is its method independence. A part of our 'Loss claim management' the business process model can be seen in Fig. 1.

There are several attributes that can and have to be set or defined when modelling a business process in ADONIS. The "skeleton" of a business process can be easily formed with activities, decision points, parallelism and merging objects, logical gateways and events, but this can be—and needs to be— detailed more.

The vertical level in detail of a business process model provides its focus point: operational areas, process areas, process models, sub-processes, detailed activities, or even deeper; the algorithms.

The horizontal level in detail of a business process model provides the level of extra information of the business process: organizational information can be specified in an organogram (working environment model in ADONIS) then the roles can be referred in the RACI matrix of the process model; the inputs and the outputs can be linked to the business process model with the IT system elements as well. If needed, key performance indicators and risk with controls can be specified for the process models too.

The decision about which levels to use from the abovementioned, and the degree of detail necessary always depends on the scope of the modelling project. A business process model is complete when it is detailed enough for proper usage.

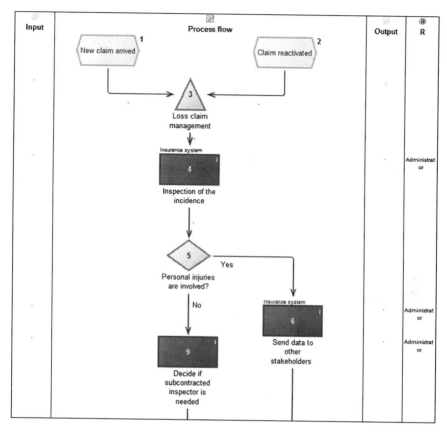

Fig. 1 The start of 'Loss claim management' process

So all projects with business process modelling have to start with the specification of the usage and needs, which gives the conventions of modelling. The book of conventions has to be known and accepted by everybody who is in the project. Based on this, everybody can model business processes in the same way, with the same degree of detail, and the models will mean the same for everybody.

In the ProKEX project business process models are used to gather knowledge from them. The following parameters have to be set to achieve this goal during the modelling of business processes:

- the logical "skeleton" of the business process model with the core objects (e.g. task, parallelism, merge, etc.);
- the organizational structure needed for the business process model, in one or more working environment models;
- the inputs and outputs needed for the business process model, in one or more document models;
- the IT elements needed for the business process model, in one or more IT system models;

- name of activities in the business process models;
- description of activities in the business process models;
- the Responsible role for all the activities in the business process models;
- input, output, IT system information for all the activities in the business process models, where available.

These parameters are no more than in an average modelling situation, so this meets average business needs. We will show, that based on the business process model, we are able to harvest the required knowledge for the business process.

2.2 Initial Modelling of Processes

The basis of our multi-lateral approach is general control-flow oriented business process models. The process modelling starts with the close observation of an existing, real-life process at the given organization. The first stage is to conduct interviews with all of the stakeholders of the process to be recorded at the company, assess already existing process documentation, and document the process development meetings and materials prepared during the actual project. A thorough inspection of the underlying IT infrastructure is also necessary.

The ever-recurring problem of capturing processes is the level of granularity. Setting this appropriate level can be thought of as an optimization problem in itself. If a process model is too superficial it will not contain enough information to draw conclusions, conduct redesign or utilize it in any other way. A modelling architecture with unnecessarily frittered details or a model with inhomogeneous granularity results in confusing process architecture, and consumes unnecessary resources to create, maintain and manage. Ternai et al. collect the parameters that have to be set in order to use a process model as a basis of semantic transformations (Ternai and Török 2011), The level of granularity in modelling a process is set to grant the ability to attach corresponding concepts, such as roles or information objects to the model.

At this point, the process structure, and meta-information for the IT and organizational viewpoints are recorded, all relevant information resources are elaborated, but organizational knowledge is unstructured, hard to identify and has various, heterogeneous sources.

2.3 Additional Modelling Layers

After finalizing the basic process flow, the specific activities within the process model have to be aligned with roles and responsibilities. We have to capture a view of the inner stakeholders of the organization. The first stage is to collect all the roles that are related to the given process and gradually examine which roles have any relation with a given activity. This task is carried out on the theoretical ground of the RACI responsibility matrix. It is necessary to determine which explicit roles are being played by which stakeholder at the level of a given activity. More precisely, we define according to the RACI which role is Responsible for the performance of the activity, which role is Accountable for it, which roles need to be Consulted during the execution of the activity, and who to be Informed about the advance, obstacles, completion or other information related to the given activity.

This knowledge is the basis of the proposed output, namely to be able to present the knowledge items required by a person in a given role, or in a broader perspective, in a given position.

There are two additional modelling dimensions that play an important part in enriching process information:

Many organizations have a well-structured IT infrastructure map, and in a higher-level process model, IT architecture elements are assigned to the process model at activity level. Modelling tools incorporate sub-models of the company's IT infrastructure. In this sub-model we define the major systems, tools or resources, which will play an active role in our processes, and associate these elements at the activity level of the process model.

Documents are also essential artifacts of business processes; various documents playing various roles are created, transferred, and utilized as a source of knowledge and information. These documents have to be taken into account throughout the complete BPM lifecycle, and in this way also incorporated into the process models.

2.4   Multilateral Process Views: Process Coupling via Semantic Transformations

The resulting complex process models contain interconnected, multilateral information in the following areas of the recorded processes:

- process structure, process hierarchy
- organizational structure, roles and responsibilities at activity level
- mapped explicit knowledge
- IT architecture
- document structure

In order to make use of this holistic process-space semantic transformations need to be applied to the models. The goal is to provide a machine-readable representation for further utilization in the form of ontologies.

Since the complex process models hold both process knowledge and domain knowledge these transformations have to be conducted respectively.

2.5   Process Ontology Creation

In this section, the focus point is the mapping of conceptual models to ontology models by using the meta-modelling approach. Meta-models provide intuitive ways of specifying modelling languages and are suitable for discussion with non-technical users. Meta-models are particularly convenient for the definition of conceptual models.

In our proposed approach, we discuss how to establish the links between model elements and ontology concepts. Ontologies basically provide semantics and they can describe both the semantics of the modelling language constructs as well as semantics of model instances (Kramler and Murzek 2006). There are three ways to create business process ontologies; reusing or extending an existing ontology; using a framework (such as the framework of SUPER[2] project); or transforming the output of a BPM tool into an ontology format. In our solution we used a process ontology we created using the output of a BPM tool, and our own mapping method.

In order to extend and map the conceptual models to ontology models, the models are exported in the structure of the ADONIS XML format. Every object from the business process model will be an 'instance' in the XML structure, the attributes have the tag 'attribute', while the connected objects (such as the performer, or the input/output data, which are stored in another model in the Adonis tool) have the tag 'interref'. A part of an XML export can be seen in Fig. 2.

The "conceptual models—ontology models" converter maps the Adonis Business Process Modelling elements to the appropriate Ontology elements at the meta-level. The model transformation aims to preserve the semantics of the business model. The structure of the business process model can be transformed with all of its objects and their attributes into the process ontology. The general rule we follow is to express each ADONIS model element as a class in the ontology and its corresponding attributes as attributes of the class. This transformation is carried out by the means of the XSLT script that performs the conversion. A sample part of the transforming XSLT code (mapping the 'Input' to an ontology element) can be seen in Fig. 3.

In order to specify the semantics of ADONIS model elements through relations to ontology concepts, the ADONIS business model must first be represented within the ontology. In regard to the representation of the business model in the ontology one can differentiate between a representation of ADONIS model language constructs and a representation of ADONIS model elements. ADONIS model language constructs such as "activity", as well as the control flow are created in the ontology as classes and properties. Subsequently, the ADONIS model elements can be represented through the instantiation of these classes and properties in the ontology.

The process ontology metamodel is based on previous results (Ternai and Török 2011), but it is extended in order to manage multiple processes in one ontology. It is as follows (Fig. 4):

- Process_stage: class, activity of the process
- Actor: class, represents a Role which is part of the RACI

---

[2] http://ip-super.org.

```
<INSTANCE id="obj.197304" class="Activity" name="Inspection of the incidence">
<ATTRIBUTE name="Position" type="STRING">NODE x:8.5cm y:8cm index:55</ATTRIBUTE>
<ATTRIBUTE name="Description" type="STRING">The first and foremost information to collect is to
decide if personal injuries are involved or not. This is vital since claims with personal injuries in
the Non-Life domain statistically results in an order of magnitude higher disbursements then other
claims, so the insurer has to conduct a particularly precautious procedure. </ATTRIBUTE>
<INTERREF name="Responsible for execution">
<IREF type="objectreference" umodeltype="Working environment model" tmodelname="Book" umodelver=""
tclassname="Role" tobjname="Administrator"></IREF>
</INTERREF>
<INTERREF name="Accountable for approving results"></INTERREF>
<INTERREF name="Cooperation/participation"></INTERREF>
<INTERREF name="To inform"></INTERREF>
<ATTRIBUTE name="RACI/DEMI visualisation" type="EXPRESSION">EXPR val:0</ATTRIBUTE>
```

Fig. 2 XML export of the business process model (fraction)

```
<xsl:for-each select="//INTERREF[@name='Input']/IREF[generate-id() = generate-id(key('IREF',
@tobjname)[1])]">
    <xsl:variable name="className" select="functx:words-to-camel-case(@tobjname)" />

    <Declaration>
        <Class>
            <xsl:attribute name="IRI">#<xsl:value-of select="$className" /></xsl:attribute>
        </Class>
    </Declaration>

    <SubClassOf>
        <Class>
            <xsl:attribute name="IRI">#<xsl:value-of select="$className" /></xsl:attribute>
        </Class>
        <Class>
            <xsl:attribute name="IRI">#Data</xsl:attribute>
        </Class>
    </SubClassOf>

</xsl:for-each>
```

Fig. 3 Fraction of XSLT code transforming 'Input' attribute to an ontology element

- IT_system, class, the supporting IT system element of the activity
- Data_object, class, inputs and outputs of the activity
- Parallel, Merge, Decision_point: classes, other objects from the process models than activity
- followed_by: relation of the Process_stage class, connects a following activity to the previous one
- performed_by: relation, connects a Process_stage with an Actor
- uses_system: relation, connects a Process_stage with an IT_system
- uses: input: relation, connects a Process_stage with a Data_object, if it is the input of the activity
- produces_output: relation, connects a Process_stage with a Data_object, if it is the output of the activity

The linkage of the ontology and the ADONIS model element instances is accomplished by the usage of properties. These properties specify the semantics

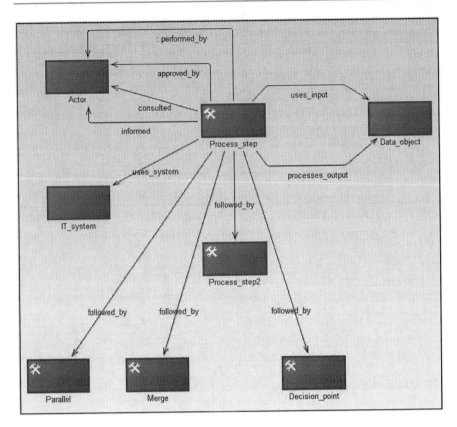

Fig. 4 The process ontology metamodel

of an ADONIS model element through a relation to an ontology instance with formal semantics defined by the ontology.

3   Case Study

In the course of the Insurance pilot of the ProKEX project we modelled close to 100 processes of an insurance company. Our project partner is a medium-sized, Hungarian insurance company operating both in the Life and Non-Life line of insurance business. The insurance company is relatively young, and founded by Hungarian stakeholders only 8 years ago. The business processes are well-grounded enough for a deeper inspection, and they are not hindered by legacy organizational fixations, but provide the necessary means for process enhancement and efficiency improvement.

For the case study we selected two complex processes that enabled us to envision the proposed solution.

3.1   Loss Claim Management

The first process is the loss claim management of the Non-Life branch. Loss claims arise either from new claims of the insured parties or from reactivating a claim when new information emerges regarding the issue. Every aspect of the issue is collected in a virtual claim issue file. The process starts with the inspection of the incident. The first and foremost information to collect relates to whether personal injuries are involved or not. This is vital since claims with personal injuries in the Non-Life domain statistically result in order of magnitude higher disbursements than other claims, so the insurer has to conduct a particularly cautious procedure. In most cases, especially if the estimated loss exceeds a given limit, an inspection or a verification of evidence is necessary. This is undertaken by subcontracted inspectors, who are the experts in the issue's insurance coverage field. This sub-process involves comprehensive support of integrated IT systems which organize the information flow between the roles played by the parties. If the amount of loss is determined, a decision mechanism within the insurer organization is triggered that results in a final decision on the claim. Throughout the process several notifications and correctional provisions might be necessary among the parties, aided by the underlying IT infrastructure.

When the insurer decides the magnitude and other conditions of the disbursement an administrative sub-process takes place involving the notification of the stakeholders of the issue, the decision on the further existence of the insurance contract, and managing the effective payment of the disbursement. If the loss results in the contract becoming obsolete, (e.g. a vehicle is deemed a total loss), the insurance contract is discontinued by the insurer, which might require further action in settling overdue or overpaid balances.

If a third party is involved, and the loss claim has been fully undertaken by the insurer, a regression process starts, that attempts to identify the insurer of the third party and negotiate based on the legal regulation or bilateral agreements between the insurers.

3.2   New Insurance Offer

The new insurance offer process was recorded for the Life insurance field-of-business of the insurance company. In many ways it can be regarded as a strongly regulated sales activity. It starts by creating a personalized offer for a prospective insured client and ends with the contract signature or the denial.

A request for a new offer always originates from an agent or representative distributor of the insurance company. The original offer documentation is prepared in one of the sales support systems. When the documentation arrives a workflow issue is created automatically with all the necessary information about the parties and the proposed life insurance contract. From this point the progress of the offer can be tracked through the workflow issue. The person responsible for the offer is a designated employee of the new-business department.

The first task is to determine the identity and the eligibility of the main parties of the offer, basically the life insured, the contracted, the beneficiary parties, and the agent eligible for commission. All available information about the parties are recorded on the issue with special care for data integrity, duplication elimination and data quality management. If any of the parties are already existing parties on other contracts of the insurer, the necessary connections have to be created, since these connections might influence the decisions on the current issue.

The agent on the offer has to be a contracted, active insurance provisioning partner of the insurer. The examination of the agent includes a thorough inspection involving a designated scoring method, including the calculation and update of the so-called "ABC indicator", which qualifies the agent based on the commission balance, the outstanding premiums of the agent's contracts, and the rate of early contract deletion. The offer issue continues on two parallel threads: health and financial risk assessment.

Based on the conditions of the offer and the regulations of the insurer, the administrator has to decide, whether it is necessary to conduct a health risk evaluation. In this case, the issue is handed over to the designated health risk assessment team. The health risk evaluation can take place simply based on the available documentation and statistical data, or it might require a medical examination of the life insured parties. If the medical examination is necessary, it has to be ordered from a third party service provider. At the end of the sub-process, the team submits a recommendation to the new business administration, where a decision is made, that in some cases includes the insurers' leading medical expert.

The term for the financial risk assessment in the insurance domain is prevention. The aim of the prevention sub-process is twofold: it stops the customer from undertaking a financial commitment that is beyond his/her financial means, and also protects the insurer from entering into a contract that is likely to fail abortively. The prevention starts with an internal evaluation of the customer, and if necessary, includes a personal interview usually conducted over the telephone. The interview itself is a workflow sub-process that leaves out the agent and directly contacts the contracted party. It ensures that there is a clear intention for the contract that all the necessary information was received, and the contractor is aware of the obligations and risks arising from the proposed life insurance contract.

If both types of risk assessment have been successfully concluded the new business department examines if all the necessary proclamations and statements have been received by the insurer. In the event that any obligatory elements are missing, the department contacts the agent or the contractor directly and requests the completion of documents. This sub-process might require multiple workflow issues. If the time interval for the completion exceeds a designated limit the offer is closed and the parties are notified.

The final inspection is conducted by two responsible team members to avoid potential abuse. Upon denial of the offer, the new business department issues official notification of the parties and closes the offer. If the final decision is positive, the offer receives an approved status. In the life insurance domain there is no prolonged payment, after the final approval the issue is an order waiting for financial settlement. When the first premium arrives to the offer, it is automatically converted to an active contract state.

3.3   Transforming a Process Model to a Process Ontology

In this section we present our approach via the two above described business processes. At the starting point there are only the two business processes modelled in Adonis Business Process Management Toolkit.

3.3.1   The Graphical View of the Process Model

In a business process model, there are objects relevant to the model and to understanding the process itself. A graphical process model has different object shapes for different parts of a process. Generally, there are tasks, gateways, lines and other objects—based on the granularity of modelling. In the Adonis BPM Toolkit, the basic object is the 'Activity'.

In Fig. 5, there is an activity from a process model. In this graphical representation the following can be seen:

- The name of the activity
- The input (left side) and the output (first lane in the right) documents
- The RACI information (other four lanes in the right)
- IT system (in the upper left corner)
- A letter 'I', indicating that there is a description written for this activity
- The number of the activity in the process model

Every object in the process model has a Notebook, where its properties can be set. Opening this Notebook, the aforementioned attributes (name, input, output, IT system) can be modified. An important attribute is the description of the activity, which is only visible in the Notebook (Fig. 6).

There are not only Activities in a process model, but Triggers, Decision points, Parallelities and Merges, as well as End events too. For our purposes Triggers are not important, but the others are.

A Decision point is in Fig. 7, with two possible following activities. This means that only one of them will be executed during the process since a decision point is an exclusive gateway.

In Fig. 8, the Parallelity and the Merge can be seen. This means that both of the activities are carried out in the process simultaneously, and when both of them are ready the process can move to the next activity following the Merge object.

3.3.2   The XML Export of a Process Model

In order to create the process ontology it is first necessary to create an XML export from the process model. The XML is a well-structured, machine readable format, therefore it is suitable for our purposes.

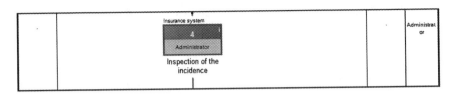

Fig. 5 Activity in the process model

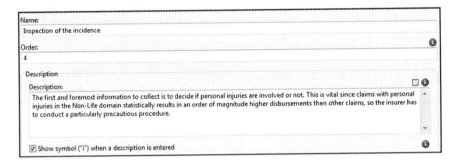

Fig. 6 Description in the Notebook

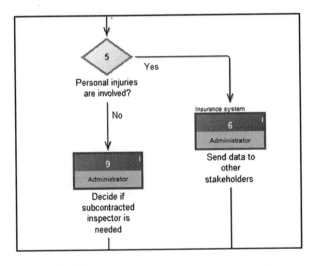

Fig. 7 Object nr. 5 is a Decision point

In Fig. 9 it can be seen that in the process model export every object has the tag <INSTANCE>, and their attributes have the tag <ATTRIBUTE>. The description is in the <ATTRIBUTE type= "Description">, as a string.

In Fig. 10 <INTERREF> tag is used instead of <ATTRIBUTE>. In a process model, when an object is stored in another model, but when we want to link it to another object, <INTERREF> tag will be used in the export. For example, in Fig. 10, for the Activity "Delegate inspector" the Document "Claim" is linked as an

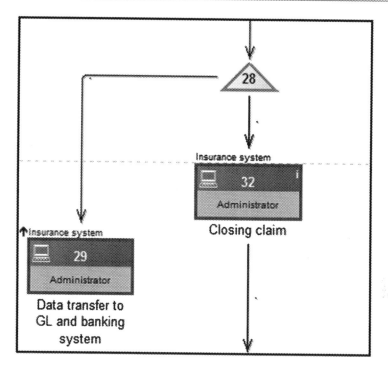

Fig. 8 Object nr. 28 is Parallelity

```
<ATTRIBUTE name="Description" type="STRING">The first and foremost information to collect is to
decide if personal injuries are involved or not. This is vital since claims with personal injuries in
the Non-Life domain statistically results in an order of magnitude higher disbursements then other
claims, so the insurer has to conduct a particularly precautious procedure. </ATTRIBUTE>
```

Fig. 9 XML export for attribute 'Description'

Fig. 10 Input and Output attributes in Notebook and in the export

Input, and the Document "Policy summary" is linked as an Output from the Document model "Insurance documents".

The same method is used for IT system elements, so for Activity "Report in IT system" the IT system "ClaimHandler" is linked from the IT system model "Insurance IT", as can be seen in Fig. 11.

Fig. 11 IT system element in Notebook and in the export

In a process model, clarifying roles and responsibilities is often carried out by a responsibility assignment matrix (RACI matrix), which describes the participation by various roles in completing tasks for a business process.

RACI are acronyms derived from the four key responsibilities most typically used: Responsible, Accountable, Consulted, and Informed.

- **Responsible**: Those who do the work to achieve the task. There is at least one role with a participation type of responsible, although others can be delegated to assist in the work required (see also RASCI below for separately identifying those who participate in a supporting role).
- **Accountable** (also **approver** or final **approving authority**): The person ultimately answerable for the correct and thorough completion of the deliverable or task, who also delegates the work to those responsible. In other words an accountable must sign off (approve) work that the person responsible provides. There must be only one accountable specified for each task or deliverable.
- **Consulted** (sometimes **counsel**): Those whose opinions are sought, typically subject matter experts; and with whom there is two-way communication.
- **Informed**: Those who are kept up-to-date on progress, often only on completion of the task or deliverable; and with whom there is just one-way communication.

IT system element in Notebook and in the export

The Notebook view of the RACI is in Fig. 12, and its export is in Fig. 13, where the <INTERREF> tags are used again, since the Roles are stored in a "Working environment model" in Adonis.

Since we want to use the process model not only as a structural definition of tasks but also as the holder of the required knowledge of each task and their responsible roles, we have ran text-mining algorithms to gather knowledge elements from the process models.

3.3.3 The Process Ontology of the 'Loss Claim Management' Process

The process ontology of the Loss claim management process is generated from the process model, via XML and XSLT transformation. The meta-model of the ontology was described above, so those classes can be seen in Fig. 14, in Protégé 5.

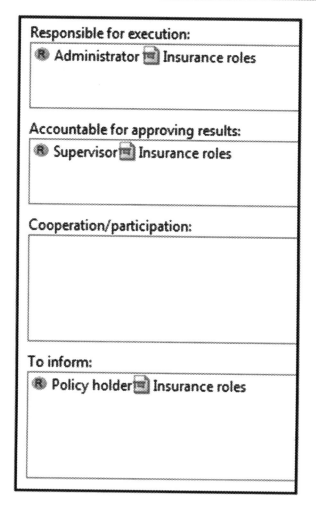

Fig. 12 RACI in Notebook

```
<INSTANCE id="obj.197271" class="Activity" name="Finalizing amount to be paid">
(...)
<INTERREF name="Responsible for execution">
<IREF type="objectreference" tmodeltype="Working environment model" tmodelname="Insurance roles"
tmodelver="" tclassname="Role" tobjname="Administrator"></IREF>
</INTERREF>
<INTERREF name="Accountable for approving results">
<IREF type="objectreference" tmodeltype="Working environment model" tmodelname="Insurance roles"
tmodelver="" tclassname="Role" tobjname="Supervisor"></IREF>
</INTERREF>
<INTERREF name="Cooperation/participation"></INTERREF>
<INTERREF name="To inform">
<IREF type="objectreference" tmodeltype="Working environment model" tmodelname="Insurance roles"
tmodelver="" tclassname="Role" tobjname="Policy holder"></IREF>
</INTERREF>
(...)
</INSTANCE>
```

Fig. 13 RACI in the export

Fig. 14 The process
ontology classes in Protégé 5

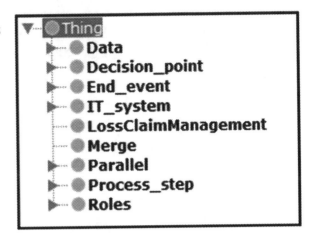

It is worth mentioning that the process itself is now a class, but it is a development issue whether creating a class 'Process' would be better in order to manage more business processes in one ontology.

In Fig. 15 all the classes are open (except the Process_stage), and their objects can be seen. These are the objects which will be linked to the activities (that are in the Process_stage class in the ontology).

As the "skeleton" of the process is formed by the activities, the most important class in the ontology is the Process_stage. In Fig. 16 the Process_stage NotificationFromDenyingClaim is detailed. We can see that this activity is followed by an End, so the process stops here if the claim is rejected. It is performed by the Administrator, and the Policyholder is informed of it. The activity has an output, the InfoLetter, and the activity itself belongs to the LossClaimManagement process.

The process ontology contains the activities of the process model as class Process_stage, decision and other logical gateways as classes Decision_point, Merge and Parallel, and—what is more important for us—the connections between these objects, so evaluating the Process_stage instances we can see the inputs and output of them, their responsible role, and description as an annotation.

Making process ontology from the process model is an innovative way of extracting knowledge from process models. In the ontology one can easily see all the tasks for one person (or role). Based on that, those tasks can be investigated more thoroughly. All the tasks have a description (since we have set this attribute as a mandatory attribute in Sect. 2.1), so the information there can be investigated with text mining methods, which we will discuss in another section of the book.

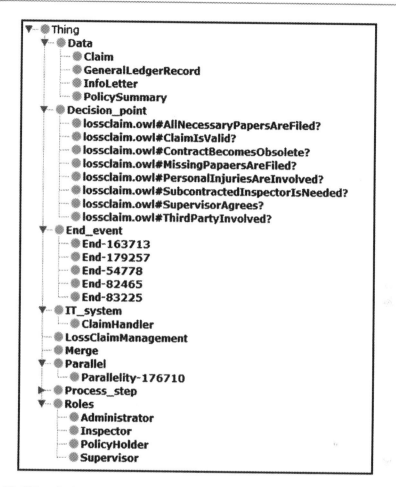

Fig. 15 Objects in the ontology classes

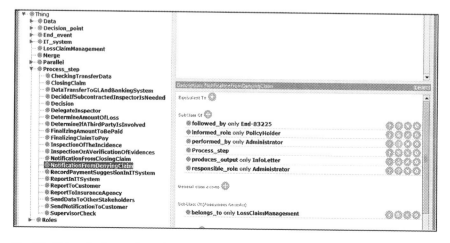

Fig. 16 Objects in the ontology classes

4   Conclusion

The chapter focused on the possible SBPM aspects of the solution utilized in the ProKEX project. We demonstrated a methodology to extract, organize and preserve knowledge embedded in business processes to enrich organizational knowledge base partways automatically. In the semantic approach, the area of knowledge necessary to complete the given process stage can be managed operationally. The solution is based on the connection between the process model and corporate knowledge base, where the process structure will be used for building up the knowledge structure. A common form of knowledge base is the ontology, which provides the conceptualization of a certain domain. We discussed how to establish the links between model elements and ontology concepts. The objective of this approach is to transform the business process into process ontology and to combine it with the knowledge base as a domain ontology in a dynamic, systematic and well-controlled solution. In the case study we illustrated the solution related to the processes of a medium-sized, Hungarian insurance company operating both in the Life and Non-Life line of insurance business.

References

Basu, A., & Blanning, R. W. (2000). A formal approach to workflow analysis. *Information Systems Research, 11*(1), 17–36.

Benjamins, V. R., Fensel, D., & Straatman, R. (1996). Assumptions of problem-solving methods and their role in knowledge engineering. In W. Wahlster (Ed.), *Proceedings ECAI-96* (pp. 408–412). Chichester: Wiley.

Berners-Lee, T., Hendler, J., & Lassila, O. (2001). The semantic web. *Scientific American, 284*(5), 34–43.

BOC Group. (2013). Business process management with Adonis. http://www.boc-group.com/products/adonis/

Celino, I., Alves de Medeiros, A. K., Zeissler, G., Oppitz, M., Facca, F., & Zoeller, S. (2007, June). Semantic business process analysis. In *Proceedings of the Workshop on Semantic Business Process and Product Lifecycle Management (SBPM-2007)*, Vol. 251, CEURWS.

Curtis, B., Kellner, M. I., & Over, J. W. (1992). Process modeling. *Communications ACM, 35*(9), 75–90.

Fensel, D., Hepp, M., Leymann, F., Bussler, C., Domingue, J., & Wahler, A. (2005). *Semantic business process management: Using semantic web services for business process management.* IEEE Conference on e-Business Engineering (ICEBE 2005), Beijing, China.

Gábor, A., & Szabó, Z. (2013). Semantic technologies in business process management. In M. Fathi (szerk.), *Integration of practice-oriented knowledge technology: Trends and prospectives* (pp. 17–28, 368 p). Berlin: Springer. ISBN: 978-3-642-34470-1.

Green, P., & Rosemann, M. (2000). Integrated process modeling. An ontological evaluation. *Information Systems, 25*, 73–87.

Harmon, P., & Hall, C. (2006). The 2006 BPM suites report. BPTrends (2006). www.bptrends.com/surveys/09-02-2006-BPMSuites-Final-Final.pdf

Hepp, M., Cardoso, J., & Lytras, M. D. (2007). *The semantic web: Real-world applications from industry.* Berlin: Springer.

Hepp, M., Leymann, F., Domingue, J., Wahler, A., & Fensel, D. (2005). Semantic business process management: A vision towards using semantic web services for business process management. In *Proceedings of the IEEE International Conference on e-Business Engineering, ICEBE 2005*.

Hepp, M., & Roman, D. (2007). An ontology framework for semantic business process management. In *Wirtschaftsinformatik Proceedings 2007* (Paper 27).

Herborn, T., & Wimmer, M. (2006). *Process ontologies facilitating interoperability in e-government, a methodological framework*. Workshop on Semantics for Business Process Management, the 3rd Semantic Web Conference, Montenegro.

Hoefferer, P. (2007). Achieving business process model interoperability using meta models and ontologies. In H. Sterle, J. Schelp, & R. Winter (Eds.), *Proceedings of the 15th European Conference on Information Systems (ECIS 2007)* (pp. 1620–1631). St. Gallen: University of St. Gallen.

Jacka, M., & Keller, P. (2009). *Business process mapping: Improving customer satisfaction* (p. 257). New York: Wiley. ISBN 0-470-44458-4.

Kalpic, B., & Bernus, P. (2006). Business process modeling through the knowledge management perspective. *Journal of Knowledge Management, 10*(3), 40–56.

Karastoyanova, D., Lessen, T., Leymann, F., Ma, Z., Nitzsche, J., Wetzstein, B., Bhiri, S., Hauswirth, M., & Zaremba, M. (2008). A reference architecture for semantic business management systems. In *Multi konferenz Wirtschaftsinformatik*. Berlin: GITO-Verlag.

Koschmider, A., & Oberweis, A. (2005). Ontology based business process description. In *Proceedings of the CAiSE* (pp. 321–333).

Kramler, G., & Murzek, M. (2006). Business process model transformation issues. http://publik.tuwien.ac.at/files/pub-inf_4629.pdf

Lautenbacher, F., & Bauer, B. (2006). *Semantic reference and business process modeling enables an automatic synthesis*. Workshop SBPM at ESWC06, Budva.

Light, A. (2005). *Gartner predicts: Nearly half of IT Jobs will be lost to automation by 2015*. UsabilityNews.com.

Pedrinaci, C., & Domingue, J. (2007). Towards an ontology for process monitoring and mining. In M. Hepp, K. Hinkelmann, D. Karagiannis, R. Klein, & N. Stojanovic (Eds.), *Proceedings of the Workshop on Semantic Business Process and Product Lifecycle Management (SBPM-2007)*. Available at: ceur-ws.org

Scheer, A.-W., Thomas, O., & Adam, O. (2005). Process modeling using event-driven process chains. In M. Dumas, W. M. P. van der Aalst, & A. H. M. ter Hofstede (Eds.), *Process-aware information systems: Bridging people and software through process technology* (pp. 119–145). Hoboken, NJ: Wiley.

Ternai, K., & Török, M. (2011). *A new approach in the development of ontology based workflow architectures*. 17th International Conference on Concurrent Enterprising—Conference Proceedings. Approaches in Concurrent Engineering. Published by: Ralf Zillekens Druck- und Werbeservice, Stolberg, Germany. ISBN: 978-3-943024-04-3. Issue Date: 20–22 June 2011.

Thomas, O., & Fellmann, M. (2007). Semantic business process management: Ontology-based process modeling using event-driven process chains. *International Journal of Interoperability in Business Information Systems, 2*, 29–43.

van der Aalst, W. M. P., Desel, J., & Oberweis, A. (2000). *Business process management: Models, techniques, and empirical studies*. Berlin: Springer.

Vernadat, F. (2002). UEML: Towards a unified enterprise modelling language. *International Journal of Production Research, 40*(17), 4309–4321. Taylor & Francis Group.

Wand, Y., & Weber, R. (1993). On the ontological expressiveness of information systems analysis and design grammars. *Journal of Information Systems, 3*(4), 217–237.

Weske, M., van der Aalst, W. M. P., & Verbeek, H. M. W. (2004). Advances in business process management. *Data and Knowledge Engineering, 50*(1), 1–8.

ProMine: A Text Mining Solution for Concept Extraction and Filtering

Saira Gillani and Andrea Kő

1 Introduction

Due to the on-going economic crisis, the management of organizational knowledge is becoming more and more important. This knowledge resides in knowledge repositories, in business processes and in employees' heads. Knowledge repositories contain explicit knowledge while employees have tacit knowledge, which is difficult to extract and codify. Business processes have explicit and tacit knowledge elements as well. Nowadays the efficiency of business processes has become one of the major motivating forces for sustainable businesses. Efficiency can be improved by increasing those people's knowledge who are involved in causing the poor efficiency of business activities. Employees' knowledge can be increased by providing the appropriate learning or training materials. However, it is difficult to ensure that the knowledge in business processes is the same as in knowledge repositories and employees' heads. Knowledge repositories have key roles in respect to knowledge management because they primarily contain the organizations' intellectual assets (this is explicit knowledge) while employees have tacit knowledge, which is difficult to extract and codify. Business processes are also important in respect to the management of organizational knowledge. The main problem that we address in this chapter is how to connect text mining to process management where both fields are different in nature since process modelling focuses on tasks (these tasks relate to each other), the flow (what comes first, what next, where they are executed in parallel) and what triggers the execution of a specific task. The nature of process modelling is *procedural* while many questions raised by modelling need answers on a contextual basis, where the context has a rather *declarative* nature (in our case the Studio ontology). The text

S. Gillani (✉) • A. Kő
Corvinus University of Budapest, Budapest, Hungary
e-mail: S.A.Gillani@ieee.org; andrea.ko@uni-corvinus.hu

© Springer International Publishing Switzerland 2016
A. Gábor, A. Kő (eds.), *Corporate Knowledge Discovery and Organizational Learning*, Knowledge Management and Organizational Learning 2,
DOI 10.1007/978-3-319-28917-5_3

mining application with its simple or sophisticated procedures bridges these two different approaches, processes the concepts and transports them to the ontology for enhancement of ontology. The purpose of ontology building and enhancement is not for its own sake, but to provide the contextual background and what is necessary for process modelling (later improvement and optimization). Therefore, the major theme of this chapter is to develop a text mining solution that extracts knowledge from processes in order to enhance or populate the existing domain ontology.

There are many solutions of information or concept extraction in literature but their goals are different. These solutions extract concepts from web or domain related documents but here in this chapter, our purpose is different, we wish to extract concepts which are related to a specific process of an organization. By doing this our final goal is to prepare a domain related ontology which covers every organizational process of that domain (in our case, the insurance domain).

Semantic similarity is used to identify concepts that have common "characteristics" and these concepts should have a minimum distance between one another. Semantic similarity detection technique can allow additional matches to be found for specific concepts not already present in knowledge bases (ontologies). It is believed that measures of semantic similarity and relatedness can improve the performance of such systems. However, these past semantic-based methods fall short in resolving the main issue: helping users to identify specific concepts related to any business process, not just the presence of domain concepts, within a relevant text. The difficulty of semantic similarity is increased when there is a reduced quantity of text like in our case where business processes do not have enough domain related data. Therefore, semantic concept extraction is still an open issue in ontology construction and there is a need to implement NLP and text mining techniques in more detail.

In this chapter, we aim to discuss a text mining solution, namely ProMine that extracts knowledge from processes. This solution helps to automatically extract new concepts in order to enhance or populate the existing ontology. The main objectives of this chapter are:

1. to introduce a text mining framework with the emphasis on its thorough semantic analysis and filtering components;
2. to propose a filtering method for concept ranking to extract the most relevant ones;
3. to propose a similarity measure for the filtering method;

The chapter is organized as follows. After the introduction Sect. 2 contains an introduction to the process of ontology learning in general. Section 3 looks at the available techniques in the computational linguistic and semantics communities including shallow and deep analysis, both at the syntactic level and the semantic level for knowledge extraction. Section 4 provides an overview of previous approaches for semantic similarity measurement and also describes their limitations. We present ProMine Ontology Learning Framework in Sect. 5, followed by the application of ProMine for the insurance domain in Sect. 6. Though a case study, the main part of ontology learning process; concept extraction, is discussed. Assessment of the ProMine application is detailed in datasets and the evaluation procedure part. Finally, Sect. 7 summarizes the chapter and discusses future work.

2 Ontology Learning

We are going to develop such a text mining solution that can extract knowledge from business processes in order to automatically or semi-automatically enhance or populate the existing domain ontology. Therefore, in this section, we will discuss an ontology learning process in general. The degree of effort that has been made in this context is therefore is discussed in this section (Fig. 1).

The most cited definition of an ontology is, "an ontology is a formal specification of a conceptualization" (Gruber 1993), while ontology learning refers to the process of creating an ontology in an automatic or semi-automatic way with limited human effort. It is also referred as a process to extract conceptual knowledge from several sources and building or creation of an ontology from scratch, enriching, or populating an existing ontology. The creation of an ontology can be represented by a touple <C, H, R, A> (Zouaq 2011) where C represents the set of classes, H represents the set of hierarchical links between the concepts, R is the set of conceptual links and A represents the set of axioms. Acquiring knowledge from a specific domain is also called ontology learning (Santoso et al. 2011). George et al. (2009) divide ontology learning into six major subtasks; term identification, synonym identification, concept identification, taxonomic relation identification, non-taxonomic relation identification, rule acquisition. Maedche and Staab described a conceptual model KAON Text-To-Onto system (Maedche and Staab 2004) that consists of four general modules of ontology learning. The first module is the ontology management component that deals with ontologies manually. The

Fig. 1 Ontology learning

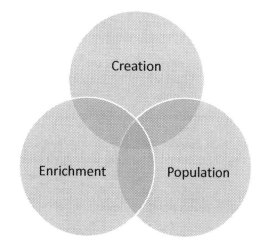

resource processing component is about preprocessing of input data that will pass to the algorithm library component which is the next component. The algorithm library component acts like a backbone of an ontology learning framework and is responsible for extraction and maintenance. The last component is the coordination component in which an ontology engineer selects the input data and chooses the method from the resource processing module and the algorithm from the algorithm library. This framework performs ontology import, extraction, pruning, and refinement. A flexible framework OntoLancs (Gacitua et al. 2008) for ontology learning is presented. This framework introduces a cyclic process that has four phases. Phase one is part-of-speech (POS) and semantic annotation phase in which domain corpus text is tagged morpho-syntactically and semantically. The second phase is extraction of concepts where a list of candidate concepts is extracted from the tagged domain corpus by applying a set of NLP and machine learning techniques. In the third domain ontology construction phase a domain lexicon is built using some outsources (WordNet, Webster) and in the last phase extracted concepts are added to a bootstrap ontology. The fourth and last phase of the framework is the domain ontology edition phase in which boot strap ontology is converted into light OWL language and then the ontology editor is used to modify/improve this domain ontology. In another study (Nie and Zhou 2008), authors placed ontology learning into three subtasks; extraction of concepts, extraction of relations and extraction of axioms. To perform these tasks they proposed an ontology learning framework OntoExtractor to construct ontologies from the corpus. The main stages of OntoExtractor are seed concept extraction, syntactic analysis, new seed concept extraction and semantic analysis based on templates. Barforush and Rahnama talked about the creation of ontologies and they described four main stages that are employed for ontology building; (i) concept learning (ii) taxonomic relation learning (iii) non-taxonomic relation learning (iv) axiom and rule learning (Barforush and Rahnama 2012).

In the same line of research, this chapter proposes a text mining solution that is based on a set of methods that contribute to all of the aforementioned major ontology learning processes (Fig. 2).

3   Ontology Extraction Tools: State of the Art

Since manual ontology construction has been costly, time-consuming and error-prone work during recent decades several semi and automatic ontology tools are presented to make ontology learning process more effective and more efficient. However, most ontology tools deal with specific ontology learning process while there are only a few tools that cover the whole ontology learning process. These tools can be broadly classified into two major categories. First those which mainly deal with plain text for ontology building while second category tools use semi structured text (Barforush and Rahnama 2012). These ontology learning tools are divided into three types by Park et al. (2010). These three parts are ontology editing

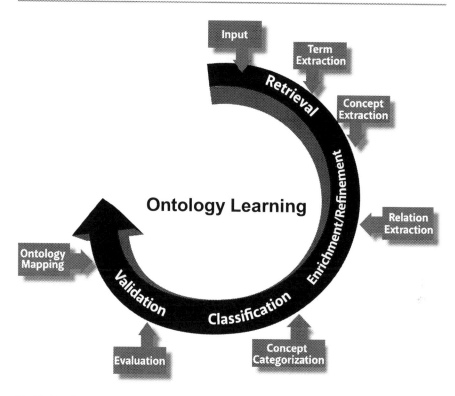

Fig. 2 Ontology learning process

tools, ontology merging tools and ontology extraction tools. Ontology editing tools provide help to the ontology engineer in acquiring, organizing, and visualizing domain knowledge. Ontology merging tools are used to make one coherent ontology from two or more existing ontologies. The third type tools are ontology extraction tools which extract concepts and/or relations by applying some NLP or machine learning techniques. In this section we will discuss some of these tools.

Though ontology editing tools (Auer 2005; Farquhar et al. 1997; Islam et al. 2010; Noy et al. 2001; Sure et al. 2002) and ontology merging tools (Noy and Musen 2003; Raunich and Rahm 2011) also reduce the ontology building time, ontology extraction tools play a more promising role in ontology automation. In this chapter our focus is on ontology extraction tools as we earlier mentioned that acquiring domain knowledge for constructing ontologies is an error prone and time-consuming task, thus, automated or semi-automated ontology extraction is necessary. In the last two decades many ontology extraction tools have been developed for this purpose.

Text2Onto (Cimiano and Völker 2005) is an ontology learning framework that is a successor (a complete redesign) of TextToOnto (Maedche and Staab 2000). Text2Onto combines machine learning and NPL techniques to extract concepts and relations. In the first phase NPL techniques, such as tokenization and sentence splitter, are applied to find an annotation set on which the POS tagger is applied and then this POS tagger assigns a syntactic category to each token. After this, machine learning and linguistic heuristics are applied to derive concepts and relations from the corpus. During this process Text2Onto applies different measures to find the relevance of a term with respect to the corpus and the results of this whole extraction process is a domain ontology. The whole process is monitored by ontology engineers. This cyclic process has some disadvantages. One of these disadvantages is the difficulty to make compound words due to lack of deep semantic analysis and due to stochastic methods Text2Onto generates very shallow and light weight ontologies (Zouaq et al. 2011). Text2Onto also lacks ontology change management and validation (Zablith 2008).

Jiang and Tan (2010) proposed a system, Concept-Relation-Concept Tuple based Ontology Learning (CRCTOL) for ontology learning. This system follows a multiple corpus based approach for key concept extraction. CRCTOL, automatically extracts semantically rich knowledge of domain related documents. To determine this, the arrangement utilizes a full text parsing technique and employs both linguistic and statistical methods to identify key concepts. The authors also proposed a rule based algorithm to discover semantic links (including both systematic and non-taxonomic links) between key concepts. An association rule mining algorithm is used for pruning unimportant links during ontology building. For evaluation they applied this system in two domains of terrorism and sports and compared the results with Text-To-Onto and Text2Onto. The results showed that ontologies built by CRCTOL are more concise and contain rich semantics as compared to other ontology learning systems. In that respect there are some limitations of this arrangement just as in other automatic learning ontology systems, and this organization also observes general concepts only and ignores whole-part relations that are likewise important in ontology building. The resulting ontology is based on domain specific documents so this ontology is not the comprehensive and accurate representation of a given domain, there is therefore a danger that such ontology will not be useful for different applications of that knowledge base. The third limitation of this system is time expensive because it performs full text parsing. To identify domain relevant concepts this system uses the term frequency measure (Domain Relevance Measure), which computes the frequency in the documents of the target area and contrasting domain documents. To achieve an accurate key concept extraction this approach involves a significant number of documents from both domains (target and contrast). However, less comprehensive domains can have a small number of relevant documents and this leads to a high skewness in key concepts and the overall performance of systems may be affected.

For ontology learning, Kang et al. (2014) introduced a novel method called CFinder, that extracts key concept for an ontology of a domain of interest. The authors described four main approaches that are oftentimes utilized for key concept extraction in literature. These are: (i) Machine learning approaches; (ii) Multiple corpus based approaches; (iii) Glossary based approaches; and (iv) Heuristic based approaches. They highlighted the problems of all these approaches such as machine learning approaches, which strongly depend on quality and the amount of training documents prior to learning. Multiple corpus based approaches (Jiang and Tan 2010) can encounter problems in performance when different domains have a different corpus size. In glossary based approaches, a set of key concepts is provided, but it is not sure that all terms of glossary carry important information from the domain because some new or too general terms may also be present in this set; so it can be hard to find key concepts of corpus on the basis of such provided terms. The writers claimed that their system overcomes all these problems. CFinder finds domain specific single-word terms that are all nouns. Hence, compound phrases were derived by using a statistical method that mixes these single words. In this process, CFinder ignores the non-adjacent noun phrases. To work out weights for these candidate key concepts of the domain, CFinder combines statistical knowledge with field specific knowledge and the inner structural pattern of these extracted candidate key concepts. This area is specific knowledge obtained from the domain specific glossary list that is furnished by the author (domain expert) or an already available glossary of that area. This list contains domain related terms. CFinder uses this list to assign a high score to domain specific key concepts. They evaluated the effectiveness of CFinder against the three state of the art methods of key extraction. The results showed that CFinder outperforms in comparison with other key extraction methods. In the beginning, the authors pointed out the drawback of glossary based approaches, but their system also uses domain specific glossary. Although in their conclusion they remarked that without domain specific cognition their system can also do well, they did not pass on any proof of this claim. In spite of its apparent limitations, key concept extraction is a major stage in ontology learning, but notwithstanding this, it is questionable how semantic links between these extracted key concepts can be estimated.

OntoCmaps (Zouaq et al. 2011) is a domain-independent unsupervised ontology learning tool that extracts deep semantic representations from unstructured text in the form of concept maps. This ontology learning tool is based on three phases: (1) a knowledge extraction phase which relies on a deep semantic analysis based on syntactic dependency patterns; (2) the integration phase builds concept maps, which are composed of terms and labeled relationships, and uses basic disambiguation techniques such as stemming, and synonym detection. These concept maps form a concept map around domain terms; and finally (3) the filtering phase where various metrics rank the items (terms and relationships) in concept maps and acts as a sieve to filter out irrelevant or overly general terms from candidates. The good thing about this ontology extraction tool is this it does not rely on any predefined template for its semantic representation and knowledge extraction is performed on each key sentence. An improvement in this work is presented by Ghadfi et al. (2014). They created a flexible language (DTPL—Dependency Tree Patterns Language) for expressing patterns as syntactic dependency trees to extract semantic relations. Through this DTPL, they extract one kind of relation from a pattern because extraction of more than one kind of relations from a pattern indicates nested patterns to differentiate by specifying dependency bindings (each dependency binding consists of a dependency link, the governor and the dependent) that should not exist when a match occurs.

In order to overcome these literature gaps in this chapter an ontology extraction tool ProMine is presented that will extract concepts from business tasks for domain ontology. To our knowledge there have been no studies carried out to address connecting text mining to process management within the context of extracting new concepts of business tasks to enrich domain ontology (defined in this chapter) for ontology learning. This proposed automatic information extraction method will comprise two basic phases: In the first phase the system will extract information from the business process and in the second phase it will enhance the extracted information using other sources such as WordNet, Wiktionary and corpus, and this enhanced information will enrich the root ontology.

4   Similarity Measures for Ontology Learning

Similarity measures determine the degree of overlap between terms or words (entities) and this measurement is based on some pre-defined factors such as statistical information about these entities or the semantic structure of these words or taxonomic relationships between these entities. The computation of the similarity between terms is at the core of ontology learning. In literature similarity measures are used for different applications of ontology learning, for example, some researchers have used similarity measures to compare the similarities between the concepts in the different ontologies, and others have used them for detecting and retrieving relevant ontologies while Saleena and Srivatsa (2015) proposed a similarity measure for adaptive e-Learning systems by comparing the concepts in cross ontology. There have been many attempts to determine similar term pairs from text corpora. It is assumed that if terms occur in a similar context then they have similar meanings (Bekkerman et al. 2001; Dagan et al. 1994). The context can be defined in diverse ways, for example, it can be represented by co-occurrence of words within grammatical relationships. Some measures of similarity are employed to assign terms into groups for discovering concepts or constructing hierarchy (Linden and Piitulainen 2004).

In this chapter our focus is on concept extraction for ontology development. We therefore see the literature related to similarity measures used for concept extraction and process in ontology development. The aforementioned ontology extraction tools extract concepts from text by using NLP or text mining techniques, and during this process many irrelevant results also come out. The majority of concept extraction approaches that are reported in literature are domain independent and few of them generally address these issues using traditional information theory metrics. In order to identify the most relevant terms it is necessary to filter out noisy data (split words or words with no meaning) and general and irrelevant terms. The output of the information extraction tool is usually a long list of words. Therefore, ranking is needed to compare several alternatives to find the best. The result of this ranking process by applying a threshold is that noisy and irrelevant words are eliminated automatically. To present adequate results to users, a filtering process is applied to the extracted knowledge. These filtering methods use different statistical and semantic measures to obtain better results. By using semantic similarity measures, important concepts and relationships (elements of domain ontology) are filtered out by comparing different candidate terms.

Researchers proposed various filtering and ranking methods based on various metrics such as co-occurrence measures, relevance measures and similarity measures to rank concepts and after ranking selecting the most relevant concepts. For term ranking, Buitelaar and Sacaleanu (2001) developed a relevance measure for information extraction. Their relevance measure is an adaptive form of standard tf.idf (Salton and Michael 1983). Their approach is task independent and completely automatic. They evaluated their method of ranking using human judgment by selecting the 100 top concepts. The results showed an 80–90 % accurate prediction of domain specific concepts. Schutz and Buitelaar (2005) developed a system (RelExt) that can be used to identify the most related pairs of concepts and relations from a domain specific text. For this purpose they used linguistic measures such as concept tagging and statistical measures such as the relevance measure (χ^2 test) and co-occurrence measure. Wang et al (2007) used entity features for filtering. From extraction method a large number of the entity pairs are generated and thus it is inefficient if they are directly classified so it is necessary to eliminate irrelevant entity pairs. (Wu and Bolivar 2008) developed an advertising keyword extraction system. This system uses the machine learning approach for ranking contextually relevant keywords. In order to model the relevance score, linear and logistic regression models are used and experiments are executed with a large set of features to obtain a keyword ranking score. Text2Onto (Cimiano and Völker 2005) relies on a distributional similarity measure to extract context vectors for instances and concepts from the text collection. In order to find the relevance of a term various measures such as the Relative Term Frequency (RTF), TFIDF (Term Frequency Inverted Document Frequency), Entropy and the C-value/NC-value are used. They also defined a Probabilistic Ontology Model (POM) that represents the results of the system by attaching a probability to them. In OntoCmaps (Zouaq et al. 2011), a set of metrics are defined to find the importance of a term such as Degree centrality, the Betweenness centrality and the Eigen-vector centrality. Betweenness is calculated by the ratio of the shortest paths between any two terms. On the basis of these metrics, the author defined a number of voting schemes to improve the precision of the terms filtering process.

Statistical measures face problems of sparsity when corpus size is small or of specialized domains. When this occurs there is a need to apply semantic measures to tackle such issues. However, it is also a difficult task to extract suitable semantic information from such a corpus. In a semantic similarity measure, two concepts are taken as input and a numeric value is returned as an output which describes how much these concepts are alike (Pedersen et al. 2007). These semantic similarity measures are used to find common characteristics between two concepts/terms. A number of semantic similarity measures have been developed in last two decades. These measures can be classified into four categories: (i) Corpus-based similarity measures, (ii) Knowledge-based similarity measures, (iii) Featured-based similarity measures and (iv) Hybrid similarity measures, as shown in Fig. 3.

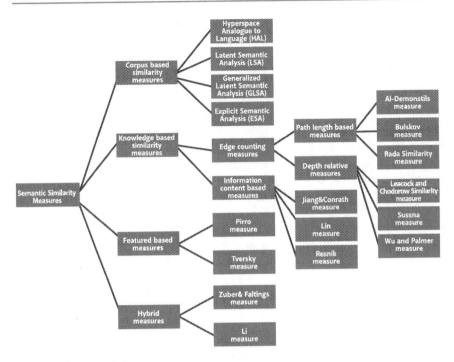

Fig. 3 Semantic similarity measures

Corpus-based measures find the similarity between concepts/terms on the basis of information that is derived from a corpus. Two well-know corpus-based similarity measures are Latent Semantic Analysis (LSA) (Guo and Diab 2012; Landauer et al. 1998) and Hyperspace Analogues to Language (HAL) model (Lund and Burgess 1996). In LSA it is assumed that words that are close in meaning occur in similar pieces of text. LSA is a high-dimensional linear association model that generates a representation of a corpus and through this representation the similarity between words is counted. In HAL, on the basis of word co-occurrences, a semantic space is created. Word ordering information (from a corpus) is also recorded in HAL.

Knowledge-based measures are used in semantic networks to measure the degree of similarity between words. Semantic networks are the networks that describe the semantic relation between words, the most famous semantic network being WordNet. In this type of networks information is in the form of graphs where nodes represent concepts and vertices represent edges. On the basis of this semantic network many similarity measures have been proposed. We can also further categorize such measures into two types: (i) edge counting measures and (ii) information content based measures. In edge counting measures, the similarity is determined by the path length measure (Euzenat and Shvaiko 2007; Nagar and Al-Mubaid 2008; Rada et al. 1989) in which the shortest path between two concepts is measured. An edge counting measure can also find similarity through depth relative measures (Qin et al. 2009; Sussna 1997; Wu and Palmer 1994) in which the depth of a particular node is calculated.

The information content is the information that a concept contains in the context in which it appears. Therefore, the main idea of information content-based measures (Formica 2008; Pirró 2009; Resnik 1995; Sánchez et al. 2011) is to use this information content of the concepts. The more common information is shared between two concepts the more similar they are to each other. If two concepts have no common information then it means they are considered maximally different. This information content can be obtained from the corpus or from a knowledge base (WordNet). An information content calculation based on WordNet performs better than corpus based information context approaches (Sánchez et al. 2011) because a sparse data problem cannot be avoided in corpus based information content similarity measures.

Hybrid measures combine the above mentioned approaches to find more accuracy. Such hybrid measures combine methods of length based measure and depth based measures. Zhou (Meng et al. 2013) has proposed a hybrid measure that combined the information content based measures and path based measures. Some researchers (Meng et al. 2013; Slimani 2013) evaluated the aforementioned measures and concluded that every semantic similarity measure has both advantages and disadvantages. Path based measures are simple to implement but local density of pair concept cannot be reflected. Information content based measures cannot reflect structural information though they are simple and effective. Hybrid measures provide more accuracy compared to other measures though these measures are more complex and also need turning of parameters.

Our proposed novel semantic similarity measure method is designed to incorporate an aggregate disease context over many patient records to create disease-specific similarity calculations. The proposed novel semantic similarity measure increases information gain from the available gene annotations.

5 ProMine Ontology Learning Framework

ProMine is an ontology extraction tool that takes input from organizational processes and extracts deep semantic representations of these organizational processes using outsources and the domain corpus. The ProMine framework basically performs two main tasks. One is knowledge extraction to extract concepts from business processes, which enriches these concepts using knowledge bases and the domain corpus. The second task is knowledge filtering to sift through the extracted information to find the most relevant concepts.

This extraction tool involves the successive application of various NLP techniques and learning algorithms for concept extraction and concept filtering.

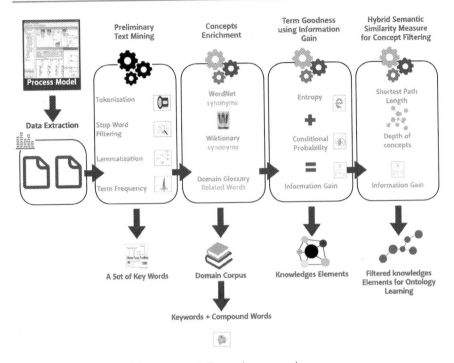

Fig. 4 ProMine: a text mining framework for ontology extraction

ProMine development is part of the ProKEX [1] project (EUREKA_HU_12-1-2012-0039) to build ontologies semi-automatically by processing the organizational processes of different domains. The initial framework of ProMine and some details of ProKEX is provided in our earlier paper (Gillani and Kő 2014). ProMine uses numerous text mining and data mining techniques for concept extraction and concept filtering which led to the development and enrichment of a domain ontology. Therefore, by using this framework an ontology can be built rather to enrich and populate an existing ontology because extraction of new concepts is mandatorily part of the whole ontology learning process.

We have developed a prototype workbench that performs the aforementioned two tasks; the knowledge element extraction and concept filtering to find the most relevant terms of a domain from the extracted knowledge elements. This prototype will show our proposed framework's efficacy as a workbench for testing and evaluating semantic concept extraction and filtering.

This section embodies ProMine as a framework to extract knowledge elements from the business processes as illustrated in Fig. 4. The workflow of our ontology framework proceeds through the phases of (i) Data extraction from organizational

[1] EUREKA_HU_12-1-2012-0039, supported by the Research and Technology Innovation Fund, New Széchenyi Plan, Hungary.

process (ii) Text preprocessing extracted data by applying natural language processing (NLP) techniques; (iii) the Concept Enrichment Phase to extract concepts from the domain corpus and other sources; (iv) it focuses on the filtering process and introduces our proposed new hybrid semantic similarity measure. Below we provide detailed descriptions of these phases.

5.1 Data Extraction

A major difference between existing ontology extraction tools and ProMine is the data extraction phase that starts from a small sized input file while in the case of already developed ontology learning tools the input is a large sized corpus or any existing ontology. ProMine's input file is actually the output file of an organizational process by using a process model. As mentioned earlier, a process can be divided into different tasks. These tasks have various attributes such as description, responsibility, execution related information (order, triggers, and events) and information about all attributes are in this input file. Our focus is on the description attribute of a task because it contains explicit and tacit knowledge elements about tasks in an embedded way. This input file is in the form of XML. In the first stage of this framework (data extraction phase), the pertinent information from this input file is extracted automatically by ProMine. After extracting specific text from the input files it is saved into text files according to all the tasks of a business process.

5.2 Preliminary Phase: Preprocessing of Data

After text extraction from the organizational processes, the most crucial part, the cleaning of extracted text starts. This preprocessing of data transforms the unstructured text in such a form that it can now be easily processed for further processing (concept extraction). This preprocessing module ensures that textual data is in such a form that text mining or data mining techniques can be applied to it to extract useful knowledge or patterns. Text preprocessing is an integral part of the natural language processing (NLP) system. Text preprocessing includes various NLP and text mining techniques such as tokenization, stop word removal, part-of-speech (POS) tagging, stemming or lemmatization and frequency count.

The main objective of this phase is to obtain some key terms and the weight of each term is based on the frequency of the term in an input file. For multivariate text analysis, in ProMine, the following preprocessing stages have been implemented.

Tokenization In this process unstructured text is segmented into discrete words that are called tokens and these words are our processing units. At this stage, word boundaries are defined and this process is totally domain dependent. There are various ways to define these boundaries. For English language text, white spaces or punctuation characters. This process is also called sentence segmentation.

Remove Stop Words Stop word filtering is applied to reduce the dimensionality of tokenized data. In this process the most common but unimportant words that have no semantic content relative to a specific domain are removed from the data. This type of data has little impact on the final results so they can be removed. The list of words is user defined so modification of it is possible. This process is applied to save storage space and to increase processing.

Part-of-Speech (POS) Tagging POS identifies lexical patterns in the text. This process helps in tokenization and it is necessary to identify valid candidate terms based on predefined POS patterns. POS removes the disambiguation between homographs, and it will also provide help in the next coming phase of concept enrichment, especially in ProMine.

Lemmatization This process is applied to extract word roots. Lemmatizer maps a token into its lexical headword or base word (lemma) as verbs are mapped to the infinitive form and nouns are mapped to the nominative singular form. This mapping transforms the word into its normalized form.

Key Term Extraction At the end of this preliminary phase a set of unique key terms will be extracted by applying a well-known statistical filter of frequency count. We set a minimum threshold for extracting the maximum important key terms.

5.3   Concept Enrichment

At the end of preliminary phase, a set of unique key words is created against each organizational task. This phase can be divided into two stages; in the first stage extracted synonyms from various lexical resources and in second stage compound words are made using the domain corpus.

A set of key words that came from the description attribute of a task from the input file may not provide enough information to generate knowledge elements for ontology enrichment because this description attribute contains little information about the task. In order to enrich the vocabulary of required knowledge elements, some language engineering tools such as WordNet (Miller 1995) and Wiktionary are used. WordNet is a semantic lexical database that contains a synset against each word and words in this synset are linked by semantic relations (Luong et al. 2012). The current version 3.0, WordNet contains 82,115 synsets for 117,798 unique nouns. The second lexical database we have used is Wiktionary, is larger in size as compared to WordNet. Like Wikipedia, any web user can edit it, which results in a rapid growth of its content. However, semantic relations in parsed Wiktionary are less than WordNet. Therefore, we have used both WordNet & Wiktionary as external resources to expand a concept's vocabulary. For every key word that has been extracted after a first phase, we acquire a set of synonyms from WordNet and

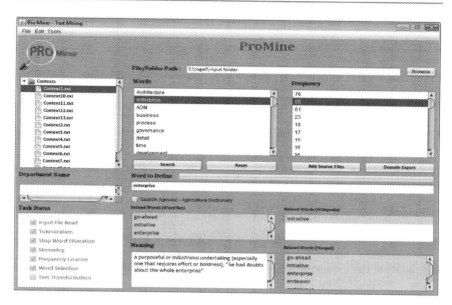

Fig. 5 ProMine: data extraction and word expansion

Wiktionary. The synonyms are the semantic variants of a given word. In ProMine there is a flexibility to add domain lexical resources, for example in one experiment on food safety domain, the AGROVOC multilingual agricultural thesaurus is also used for obtaining more domain related concepts. At the end of this stage a combined list of synset is produced against each key word (Fig. 5).

In the next stage of concept enrichment phase, a domain corpus is used. WordNet and Wiktionary are not domain dependent lexical databases. WordNet has different senses of a word so it is always possible that many irrelevant words (that semantically have the same meaning, but which are not part of specific domain) may also be generated. There is a need to eliminate such words from this synonym list. For this purpose, a domain corpus is used that includes domain glossaries, legal documents, any domain related published or unpublished documents. ProMine, prepares this corpus by itself. It takes different format (pdf, word, ppt) files and transforms them into a text file. After the transformation, preprocessing techniques which are described in the previous section are applied to this domain corpus. Now, a procedure of a few stages is applied to this preprocessed domain corpus to filter out the above mentioned ambiguities. An important function of this procedure is to extract a set of domain-specific key-concepts automatically in the form of compound words. Concepts can be more informative in compound or multi-word terms as compared to single words. However, WordNet database provides only a few compound words/multiword terms. Therefore, at this stage, multiword terms are also stretched from the given corpus because these multiword terms represent concepts that are more important to acquire meaningful knowledge elements. The resulting candidate words from the first stage of the concept enrichment phase are passed through the procedure described as follows.

1. As preprocessing has been applied to the corpus. Two-word noun compounds (bigram) via the POS tags are extracted from the corpus. The noun—noun compound is a common type of multiword expression in English. From these two-word noun compounds one word is our candidate word from a candidate list of words and the other word is from the corpus. We pass every candidate word in the corpus and if a noun is found either its right or left is joined to the candidate word to make a compound (bigram) word. If no noun word is found on the right side or left side of the candidate word it keeps it as a single word (unigram). During joining it is noted that if nouns are separated by full stops or commas (punctuation marks) then the system will not join two such nouns.

2. Once the compound words are identified automatically the next stage is to count the frequency of all the words including unigrams and bigrams. If any candidate word does not occur in the corpus or its recurrence is below a defined threshold this word will be dropped from the list. In this way all irrelevant words from the list of synonyms are also dropped because if some synonyms are not found in the corpus they are automatically eliminated from the output list. If any compound word is below the threshold, then our system will check the other content word (not candidate word) and if it passes the frequency threshold it will then remain in the list, but if the second content word does not pass the frequency threshold it will be removed from the list.

3. As a result of this phase, a rich list of concepts against each key word will be generated.

We also did a trigram compound word experiment but it didn't bring any valuable information. We already get more information with the two word nouns (bigram) selection.

5.4   Concept Filtering Based on Semantic Similarity Measure

Until the last phase unrelated terms (conceptually, not related to a specific domain) from a set of synonyms terms (from WordNet & Wiktionary) of a given key term were removed. However, the resultant word list consists of lexical terms which are hundreds in number. This high dimensionality of the feature space is the major particularity of text domain. The unique concepts or potential concepts are considered as feature space, these lists of concepts can be considered as high dimensional and sparse vectors. At this stage in our proposed framework we are reducing the feature space by selecting more informative concepts from this concept list by using a concept filtering method. Conventionally, in most ontology learning tools, statistical measures such as TF-IDF, RTF, entropy or probability methods are used for the filtering process (Cimiano and Völker 2005). To identify important lexical terms, ProMine used an innovative approach, which is a combination of statistical and semantical measures. We have proposed a new hybrid semantic measure. This module consists of two phases; in one phase for each candidate concept its information gain (IG) is calculated by using the domain corpus and in the second phase—to find more semantically representative candidate concepts—we used knowledge bases similarity measures. Finally, a hybrid similarity measure was proposed to identify relevant ontological structures for a given organizational process.

5.4.1 Statistical Syntactic Measure (Information Gain)

ProMine uses Information Gain as a term of goodness criterion (Yang and Pedersen 1997). We find out the IG for all potential terms. First, we calculate entropy, which is the measure of unpredictability and provide the foundation of IG. Entropy is defined as

$$Entropy = -\sum_{i=1}^{m} P_r(c_i)logPr(c_i) \tag{1}$$

Where $\{c_i\}_{i=1}^{m}$ is the set of words in the target space (synonym set of key word).

After calculating entropy, we have to find out probability with respect to candidate concept by following equation

$$P_r(t)\sum_{i=1}^{m} P_r(c_i|t)logPr(c_i|t) \tag{2}$$

Where $P_r(t)$ represents candidate concept.

The information gain (IG) can now be worked using Eqs. (1) and (2). On the basis of the information gain (IG) every candidate concept is ranked and the concepts with the lowest information gain will be removed by defining a threshold value. The information gain $IG(t)$ of a candidate concept with respect to the key term is defined as

$$IG(t) = -\sum_{i=1}^{m} P_r(c_i)logPr(c_i) + P_r(t)\sum_{i=1}^{m} P_r(c_i|t)logPr(c_i|t) \tag{3}$$

At the end of this step, we have information gain for all the candidate concepts. This information gain of each candidate is used in our proposed hybrid similarity measure.

5.4.2 A New Hybrid Semantic Similarity Measure

We mentioned that at the end of the concept enrichment module processing, a list of candidate concepts will be derived against each keyword extracted from the organizational process. To find more relevant terms for our domain ontology we have presented an innovative hybrid similarity measure. The main idea of this semantic measure is that the similarity between two concepts $c1$ and $c2$ is a function of the attributes path length, depth and information gain (IG) as follows:

$$Similarity(c1, c2) = f(len, \ depth, \ IG) \tag{4}$$

Where, *len* is the conceptual distance between two nodes $(c1, c2)$ which is also known as the shortest path length between $c1$ and $c2$.

depth is the depth of concept nodes

IG is the information gain of $c1$ and $c2$

We assume that Eq. (4) can be rewritten using three independent functions as:

$$Similarity(c1, c2) = f(f1(len), f2(depth), f3(IG)) \tag{5}$$

Path length and depth is calculated from lexical database WordNet while IG is derived from the domain corpus as mentioned in Sect. 5.4.1. The details of these $f1, f2$ and $f3$ are as follow:

Path Length Attribute The conceptual distance between two concepts is proportional to the number of edges separating the two concepts in the hierarchy.

$$f1_{length}(c1, c2) = (2*deep\_max - len(c1, c2))/2*deep\_max \tag{6}$$

$len(c1, c2)$ is the length of the shortest path from $c1$ to $c2$ and

deep_max is the maximum depth of the semantic hierarchy and $2 * deep\_max$ is the maximum value that $f1(c1, c2)$ can get.

Depth Attribute Depth is another factor that affects the similarity between words. As we know, concepts at upper layers of the hierarchy in semantic networks (WordNet) have more general semantics and less similarity, while concepts at lower layers have more concrete semantics and stronger similarity. This shows the importance of the depth attribute for finding the similarity between concepts.

$$f2_{depth}(c1, c2) = \frac{2 \times depth(LCS(c1, c2))}{depth(c1) + depth(c2) + 2 \times depth(LCS(c1, c2))} \tag{7}$$

depth $(c1)$ is the length of the path to $c1$ from the global root entity in the hierarchy

$LCS(c1, c2)$ is the lowest common subsume of $c1$ and $c2$.

Information Gain (IG) This is the third attribute of our similarity measure and has already been discussed in Sect. 5.4.1.

$$IG(t) = -\sum_{i=1}^{m} P_r(c_i)logPr(c_i) + P_r(t)\sum_{i=1}^{m} P_r(c_i|t)logPr(c_i|t) \tag{8}$$

By adding these three factors our similarity function will filter out important terms, which are the potential concepts for the domain ontology. Here we called them knowledge elements.

For evaluation we have taken a case study of the insurance domain described in the next section.

6 ProMine Case: Insurance Product Development

The following is a case study of Insurance Product Development. The insurance business is heavily dependent on product development. They always have to come up with new products to attract the special needs of the customers and the insurance company has to respond to the changing demands of its customers. As mentioned earlier, this research is a part of ProKEX project and the main goal of ProKEX is to develop a domain ontology. The overriding emphasis of this ontology is on the actual flow in business endless product design.

Text mining can start as a roam scratch, even when nothing is known about the domain. In this case it is just a matter of reading down and finding concepts from the text. In other cases a seed ontology or knowledge base already exists as prior knowledge. Text mining is somehow controlled by prior knowledge. We have a basic idea of what we are looking for; this is one of the reasons why we have selected this domain as a use case.

7 Evaluation

In our Insurance Product Development use case an XML file is generated from the process model in which descriptions of various tasks of the "product development" process are defined. ProMine takes this XML as input data and its first data extracting module extracts text from the description attribute of various tasks. This description attribute contains some descriptive information about the task. After extraction the data extraction module saves this extracted text into different text files according to the tasks to be performed. After extracting this text the preliminary Phase of ProMine starts in which various preprocessing techniques are applied to this unstructured text as mentioned in Sect. 5. The output of this phase is a set of unique key words against each task. Now the main processing of our concept extraction tool, ProMine, starts. First of all, we selected "insurance" as a key word and passed it to a concept enrichment module that performs a two-phase process. In first phase a set of synonyms from WordNet as well from Wiktionary are extracted in order to find more information elements related to the insurance product development process. For example, for the keyword "insurance" a synonym list with the following elements: [policy, insurance policy, indemnity]. In the second phase, to make this list richer and domain related, each word including a key word, is passed through the domain corpus where compound words are compiled according to a procedure elaborated in Sect. 5.3. The result of this phase is a huge list of concepts, which is extracted against the "insurance" key word. To filter more relevant terms we applied the proposed filtering mechanism described in Sect. 5. Finally, we have a filtered list of concepts ready to add to the ontology.

After extracting these concepts list, for the evaluation of results, a domain expert was brought in qualified to rate the domain coverage of an ontology by a domain expert. The domain expert checked these new concepts and categorized them according to seed ontology classes. Experts are involved to judge whether the suggested concepts that the system ProMine produced were correct or not. The resulting taxonomy from our extracted concepts is provided in Fig. 6. Table 1 shows the number of concepts that were extracted and evaluated by the domain expert. These extracted concepts are evaluated in two categories; "Accepted" and "Rejected". Accepted is further divided into two categories; "Important", which fit into the categories of seed ontology and "Understandable" which are understandable and can be the part of ontology as they add more classes or categories in seed ontology. Rejected concepts are considered as invalid concepts.

The results show that more than 70 % of the concepts were accepted by the domain expert. We selected the insurance domain because we have some prior knowledge in the form of seed ontology and want to enrich this ontology with new concepts. However, because of this prior knowledge we found that some of the concepts are easily categorized into the seed ontology because these fit into the existing classes, however, we encountered difficulties when trying to categorize other concepts (understandably). These were concepts which belong to the policy attribute or the product development. However, on the basis of these "understandable" concepts ontology engineers can add more categories in ontology. Thus, our ontology extraction tool ProMine can be of great help in both ontology population and enrichment.

8   Conclusion

This chapter presented an ontology extraction and filtering tool for ontology learning. Our developed tool ProMine also addresses ranking and filtering relevant terms by using a new hybrid similarity measure. The novelty of this extracting tool is that (1) it extracts concepts from very little knowledge embedded in the organizational processes and using outsources enriches this knowledge and extracts a huge number of new concepts automatically without human interaction; (2) its filtering approach uses deep syntactic and semantic analysis to filter important concepts. The other contribution is that we have proposed a new hybrid similarity measure that can be used for other applications of artificial intelligence, psychology and cognitive science. The running example described in Sect. 6, gives a step-by-step demonstration of ProMine's functionality. It illustrates how and what each module contributes to the working system. Our results demonstrated that a human expert agreed with a very large proportion of the suggested concepts that ProMine

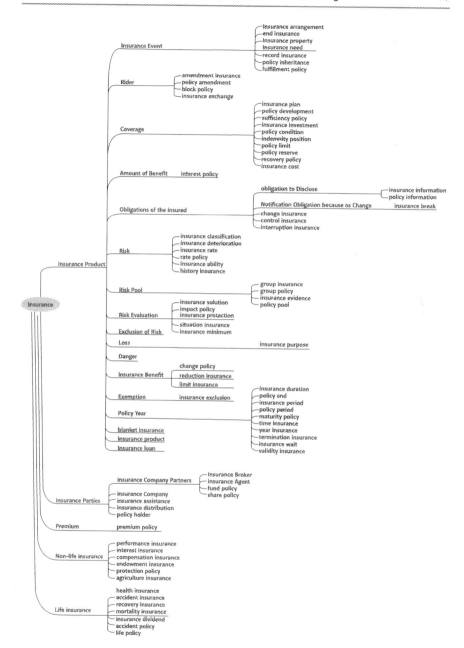

Fig. 6 Manual categorization of extracted knowledge elements

Table 1 Evaluation of constructed concepts

Concepts	Accepted			Rejected
	Important (1)	Understandable (2)	(1) + (2)	
323	162	85	247	76

produced. Many new concepts were successfully extracted and later used for the ontology population as shown in Fig. 6.

In the future we plan to test our tool ProMine testing in other domains too, and we will focus particularly on automated ontology evaluation. We will also compare our results with some other state of the art ontology extraction and filtering approaches.

References

Auer, S. (2005). *Powl–a web based platform for collaborative semantic web development.* Paper presented at the Proceedings of the Workshop Scripting for the Semantic Web.

Barforush, A. A., & Rahnama, A. (2012). Ontology learning: Revisted. *Journal of Web Engineering, 11*(4), 269–289.

Bekkerman, R., El-Yaniv, R., Tishby, N., & Winter, Y. (2001). *On feature distributional clustering for text categorization.* Paper presented at the Proceedings of the 24th Annual International ACM SIGIR Conference on Research and Development in Information Retrieval.

Buitelaar, P., & Sacaleanu, B. (2001). *Ranking and selecting synsets by domain relevance.* Paper presented at the Proceedings of WordNet and Other Lexical Resources: Applications, Extensions and Customizations, NAACL 2001 Workshop.

Cimiano, P., & Völker, J. (2005). *Text2Onto. Natural language processing and information systems.* Paper presented at the 10th International Conference on Applications of Natural Language to Information Systems, NLDB 2005, Alicante, Spain, June 15–17, 2005. Proceedings, of Lecture Notes in Computer Science (Edited by: Montoyo A, Muñoz R, Métais E).

Dagan, I., Pereira, F., & Lee, L. (1994). *Similarity-based estimation of word cooccurrence probabilities.* Paper presented at the Proceedings of the 32nd annual meeting on Association for Computational Linguistics.

Euzenat, J., & Shvaiko, P. (2007). *Ontology matching* (Vol. 333). Berlin: Springer.

Farquhar, A., Fikes, R., & Rice, J. (1997). The ontolingua server: A tool for collaborative ontology construction. *International Journal of Human-Computer Studies, 46*(6), 707–727.

Formica, A. (2008). Concept similarity in formal concept analysis: An information content approach. *Knowledge-Based Systems, 21*(1), 80–87.

Gacitua, R., Sawyer, P., & Rayson, P. (2008). A flexible framework to experiment with ontology learning techniques. *Knowledge-Based Systems, 21*(3), 192–199.

George, P., Vangelis, K., Anastasia, K., Georgios, P., & Constantine, S. D. (2009, June). Semi-automated ontology learning: The boemie approach. In *Proceedings of the First ESWC Workshop on Inductive Reasoning and Machine Learning on the Semantic Web, Heraklion, Greece.*

Ghadfi, S., Béchet, N., & Berio, G. (2014). *Building ontologies from textual resources: A pattern based improvement using deep linguistic information.* Paper presented at the Proceedings of the 5th Workshop on Ontology and Semantic Web Patterns (WOP2014), Riva del Garda, Italy.

Gillani, S. A., & Kő, A. (2014). Process-based knowledge extraction in a public authority: A text mining approach. In *Electronic government and the information systems perspective* (pp. 91–103). Cham: Springer International Publishing.

Gruber, T. R. (1993). A translation approach to portable ontology specifications. *Knowledge Acquisition, 5*(2), 199–220.

Guo, W., & Diab, M. (2012). *A simple unsupervised latent semantics based approach for sentence similarity.* Paper presented at the Proceedings of the First Joint Conference on Lexical and Computational Semantics-Volume 1: Proceedings of the Main Conference and the Shared Task, and Volume 2: Proceedings of the Sixth International Workshop on Semantic Evaluation.

Islam, N., Siddiqui, M. S., & Shaikh, Z. (2010). *TODE: A Dot Net based tool for ontology development and editing.* Paper presented at the 2nd International Conference on Computer Engineering and Technology (ICCET).

Jiang, X., & Tan, A. H. (2010). CRCTOL: A semantic-based domain ontology learning system. *Journal of the American Society for Information Science and Technology, 61*(1), 150–168.

Kang, Y.-B., Haghighi, P. D., & Burstein, F. (2014). CFinder: An intelligent key concept finder from text for ontology development. *Expert Systems with Applications, 41*(9), 4494–4504.

Landauer, T. K., Foltz, P. W., & Laham, D. (1998). An introduction to latent semantic analysis. *Discourse Processes, 25*(2–3), 259–284.

Lindén, K., & Piitulainen, J. O. (2004). *Discovering synonyms and other related words.* Paper presented at the Proceedings of COLING 2004 CompuTerm 2004: 3rd International Workshop on Computational Terminology.

Lund, K., & Burgess, C. (1996, April). Hyperspace analogue to language (HAL): A general model semantic representation. *Brain and Cognition, 30*(3), 5–5. 525 B ST, STE 1900, San Diego, CA 92101-4495: Academic press Inc JNL-COMP Subscriptions.

Luong, H., Wang, Q., & Gauch, S. (2012). *Ontology learning using word net lexical expansion and text mining.* INTECH Open Access Publisher.

Maedche, A., & Staab, S. (2000). *The text-to-onto ontology learning environment.* Paper presented at the Software Demonstration at ICCS-2000-Eight International Conference on Conceptual Structures.

Maedche, A., & Staab, S. (2004). Ontology learning. In *Handbook on ontologies* (pp. 173–190). Berlin Heidelberg: Springer.

Meng, L., Huang, R., & Gu, J. (2013). A review of semantic similarity measures in wordnet. *International Journal of Hybrid Information Technology, 6*(1), 1–12.

Miller, G. A. (1995). WordNet: A lexical database for English. *Communications of the ACM, 38* (11), 39–41.

Nagar, A., & Al-Mubaid, H. (2008). *A new path length measure based on go for gene similarity with evaluation using sgd pathways.* Paper presented at the 21st IEEE International Symposium on Computer-Based Medical Systems, 2008. CBMS'08.

Nie, X., & Zhou, J. (2008). *A domain adaptive ontology learning framework.* Paper presented at the IEEE International Conference on Networking, Sensing and Control, 2008. ICNSC 2008.

Noy, N. F., & Musen, M. A. (2003). The PROMPT suite: Interactive tools for ontology merging and mapping. *International Journal of Human-Computer Studies, 59*(6), 983–1024.

Noy, N. F., Sintek, M., Decker, S., Crubézy, M., Fergerson, R. W., & Musen, M. A. (2001). Creating semantic web contents with protege-2000. *IEEE Intelligent Systems, 16*(2), 60–71.

Park, J., Cho, W., & Rho, S. (2010). Evaluating ontology extraction tools using a comprehensive evaluation framework. *Data and Knowledge Engineering, 69*(10), 1043–1061.

Pedersen, T., Pakhomov, S. V., Patwardhan, S., & Chute, C. G. (2007). Measures of semantic similarity and relatedness in the biomedical domain. *Journal of Biomedical Informatics, 40*(3), 288–299.

Pirró, G. (2009). A semantic similarity metric combining features and intrinsic information content. *Data and Knowledge Engineering, 68*(11), 1289–1308.

Qin, P., Lu, Z., Yan, Y., & Wu, F. (2009). *A new measure of word semantic similarity based on wordnet hierarchy and dag theory.* Paper presented at the International Conference on Web Information Systems and Mining, 2009. WISM 2009.

Rada, R., Mili, H., Bicknell, E., & Blettner, M. (1989). Development and application of a metric on semantic nets. *IEEE Transactions on Systems, Man and Cybernetics, 19*(1), 17–30.

Raunich, S., & Rahm, E. (2011). *ATOM: Automatic target-driven ontology merging*. Paper presented at the IEEE 27th International Conference on Data Engineering (ICDE), 2011.

Resnik, P. (1995, August 20–25). Using information content to evaluate semantic similarity in a taxonomy. In *Proceedings of the 14th International Joint Conference on Artificial Intelligence* (pp. 448–453). Montreal, QC, Canada.

Saleena, B., & Srivatsa, S. (2015). Using concept similarity in cross ontology for adaptive e-Learning systems. *Journal of King Saud University-Computer and Information Sciences, 27*(1), 1–12.

Salton, G., & Michael, J. (1983). *Introduction to modern information retrieval* (pp. 24–51). New York: McGraw-Hill.

Sánchez, D., Batet, M., & Isern, D. (2011). Ontology-based information content computation. *Knowledge-Based Systems, 24*(2), 297–303.

Santoso, H. A., Haw, S.-C., & Abdul-Mehdi, Z. T. (2011). Ontology extraction from relational database: Concept hierarchy as background knowledge. *Knowledge-Based Systems, 24*(3), 457–464.

Schutz, A., & Buitelaar, P. (2005). Relext: A tool for relation extraction from text in ontology extension. In *The semantic web–ISWC 2005* (pp. 593–606). Berlin Heidelberg: Springer.

Slimani, T. (2013). Description and evaluation of semantic similarity measures approaches. *International Journal of Computer Applications, 80*(10), 0975–8887.

Sure, Y., Angele, J., & Staab, S. (2002). OntoEdit: Guiding ontology development by methodology and inferencing. In *Proceedings of the International Conference on Ontologies, Databases and Applications of SEmantics ODBASE 2002*. Irvine, CA: University of California.

Sussna, M. J. (1997). Text retrieval using inference in semantic metanetworks.

Wang, G., Yu, Y., & Zhu, H. (2007). *Pore: Positive-only relation extraction from wikipedia text*. Berlin: Springer.

Wu, X., & Bolivar, A. (2008). *Keyword extraction for contextual advertisement*. Paper presented at the Proceedings of the 17th International Conference on World Wide Web.

Wu, Z., & Palmer, M. (1994). *Verbs semantics and lexical selection*. Paper presented at the Proceedings of the 32nd Annual Meeting on Association for Computational Linguistics.

Yang, Y., & Pedersen, J. O. (1997). *A comparative study on feature selection in text categorization*. Paper presented at the ICML.

Zablith, F. (2008). *Dynamic ontology evolution*. International Semantic Web Conference (ISWC) Doctoral Consortium, Karlsruhe, Germany.

Zouaq, A. (2011). An overview of shallow and deep natural language processing for ontology learning. *Ontology Learning and Knowledge Discovery Using the Web: Challenges and Recent Advances, 2*, 16–37.

Zouaq, A., Gasevic, D., & Hatala, M. (2011). Towards open ontology learning and filtering. *Information Systems, 36*(7), 1064–1081.

STUDIO: Ontology-Centric Knowledge-Based System

Réka Vas

1   Introduction

Why is STUDIO a knowledge based system? How can STUDIO make an organization more knowledgeable? Knowledge-based systems (KBSs) have been an important topic in research for quite some time. The literature defines these systems in many different ways. The simplest definitions (such as Laudon & Laudon 1997[1]) describe KBSs as organizational information systems that could provide help in managing the knowledge assets of the organization. These definitions, however, are too general, since any information systems used for handling knowledge (e.g.: expert systems, data warehouses, group decision support systems or intranets) are included in the class.

Another set of definitions focus on the architecture of KBSs (Lucas & van der Gaag 1991; Akerkar & Sajja 2009). Usually three major components are distinguished: a *knowledge base* that is a repository of formal knowledge, an *inference engine* that defines the ways how the formal knowledge may be put to use and a *user interface* where "how" and "why" questions are asked. In some cases additional components are added to the above listed ones that provide instruments for filling the knowledge base, support the explanation and reasoning of decisions or enable self-learning for users (Akerkar & Sajja 2009). In practice, however, it is difficult to separate the aspects of an inference engine and knowledge base—just like in the case of STUDIO—that may hamper understanding the role and capabilities of the system

---

[1] In later versions of their book they have used the expression "knowledge management system" instead of "knowledge-based system".

R. Vas (✉)
Corvinus University of Budapest, Budapest, Hungary
e-mail: reka.vas@uni-corvinus.hu

© Springer International Publishing Switzerland 2016
A. Gábor, A. Kő (eds.), *Corporate Knowledge Discovery and Organizational Learning*, Knowledge Management and Organizational Learning 2,
DOI 10.1007/978-3-319-28917-5_4

in question. In STUDIO the knowledge is stored in the form of a domain ontology that either has inherent reasoning capabilities or may provide a base for independent applications with inference options.

In addition to these definitions several works exist that focus on the knowledge modelling aspects of KBSs. More precisely, the emphasis is placed on finding a way of formal knowledge representation. The most widely used and universally accepted techniques are logical representation, production rule representation, frame representation and semantic networks. The major advantage of defining the KBS as an outcome of a knowledge modelling process is that attention is directed to the identification of *which elements of the organizational knowledge can validly be described* in any of the formalisms for knowledge representation (Hendriks & Vriens 1999). In other words, in the course of knowledge modelling, it has to be established which part of the organizational knowledge can be identified in a formal schemata (and the kind of questions that can be answered using this formal knowledge model). An additional advantage of this approach—besides the formalization of organizational knowledge—is that efforts could be more effectively coordinated in order to explore all potential functionality of the KBS. STUDIO, on one hand provides a framework for the formal representation of knowledge in the form of a domain ontology. On the other hand, based on the context given by the formalized knowledge STUDIO also supports the design and implementation of various knowledge based applications (e.g.: adaptive knowledge testing, learning style detection, human resource preselection, etc.).

Why does STUDIO use ontologies for knowledge representation? Ontology as a tool of artificial intelligence, knowledge management, and a theoretical tool of database modelling, attempts to describe the world on a conceptual level. According to the most quoted definition "ontology is a formal explicit specification of a shared conceptualization" (Gruber 1993 p 199). That is, an ontology states knowledge explicitly to make it accessible for machines; determines knowledge only of a particular domain of interest[2] in a conceptual way applying symbols that represent concepts and their relations. While shared means that there is a consensus concerning all elements of the conceptual model. Corcho and his colleagues—based on Gruber's definition—have constructed a more precise and applicable definition: "ontologies aim to capture consensual knowledge in a generic and formal way, so that they may be reused and shared across applications (software) and by groups of people. Ontologies are usually built cooperatively by a group of people in different locations" (Corcho et al. 2003, p. 44). In other words by developing uniform conceptualizations of the domains of interest, ontologies have *consensus generating power* enabling efficient cooperation even on the organizational level. Besides knowledge sharing, ontologies also play an important role in *keeping accessible knowledge up-to-date* and in enhancing its reuse. Furthermore, through the formalization of ontologies, semantic communication and

---

[2] In other words the ontology is specified.

co-operation becomes possible not only among humans, but computers as well, enabling the efficient development and maintenance of knowledge-based systems.

Knowledge plays a vital role both in performing day-to-day activities and in reflecting on these daily routines in all organizations. At the same time, the relevance of knowledge (and/or its elements) may differ even between organizational levels and may also change over time. It is also risky to assume that the right knowledge is naturally at the right place and our knowledge workers have all the necessary knowledge at their disposal all the time. Therefore, the need for effective knowledge management tools that enable the creation, application, reuse and evaluation of knowledge is permanently increasing. In this paper we present our work in designing the *STUDIO ontology-centric knowledge-based system for effective knowledge management and personalized learning*. The ontology-based domain models are at the core of the system as they drive the creation, storage, validation and search for relevant knowledge elements.

In our architecture either business process descriptions or training materials can be applied to identify relevant knowledge elements that have to be maintained and enforced in a logical structure using ontologies. Based on the ontology-centric architecture a repository of knowledge related content—including both learning materials and test questions—have been developed to support the implementation of knowledge assessment and personalized learning applications. Our aim is to provide efficient and flexible knowledge repository functionality for supporting knowledge testing in multiple situations (such as preselection or self-assessment) and provide a mechanism for creating and enriching ontological descriptions from various sources (e.g.: business process descriptions) that enhance the storage, distribution and publishing of stored knowledge in a reusable fashion. This chapter provides a detailed description of the ontology-centric architecture and multiple application scenarios of the STUDIO knowledge-based system.

2 The Architecture of the Ontology-Centric STUDIO System

For any knowledge-based system a number of requirements need to be satisfied in order to enable the development of multiple knowledge-based applications. These requirements are the following:

- **Knowledge representation** languages that are responsible for expressing the structure of the given application have to be selected with care.
- **Knowledge organization** tools that allow for the efficient handling of even large and complex knowledge structures are also necessary.
- **Environments** that enable users to create, maintain and query knowledge are also a must in these systems (Jarke et al. 1989).

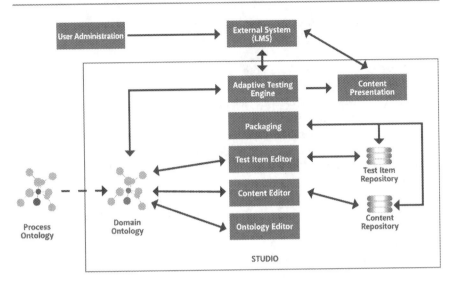

Fig. 1 Architecture of the Ontology-centric STUDIO System

Ontology-centric architecture can satisfy these requirements effectively and efficiently. The most distinguishing characteristic of the STUDIO system is the central role that ontologies play. In our architecture, knowledge is organized in a logical, multi-relational ontology structure, defined either according to business process models or learning materials representing the domain of interest.

The high-level design of this ontology-centric architecture takes a modular approach, as can be seen in Fig. 1. Besides providing the formal description of the domain of interest, the *Domain Ontology* serves as a basis for the *Adaptive Knowledge Testing Engine,* that is the primary application of the system. The structure of content is also determined by the ontology in STUDIO, meaning that every piece of content (a learning material or a test question) is connected to one (and only one) specific concept of the Domain Ontology. Learning materials are stored in the *Content Repository*, while test questions are stored in the *Test Item Repository*. Additionally, editing of the various components (domain ontology concepts, learning materials and test questions) is enabled by the respective modules of the system, namely the *Ontology Editor*, the *Content Editor* and *Test Item Editor*. The *Packaging* component enables power users and/or domain experts to develop customized scenarios for knowledge assessment by selecting certain concepts (and concept trees) from the overall domain ontology that best describe the targeted sub-domain. Finally, the *Content Presentation* module is entitled to present and visualize the stored content pieces (adaptive tests, test results, ontology visualization and learning materials) to the end users. These components will be further discussed in the following sections.

The STUDIO system has been designed to enable the flexible use of its functionality, independently from its form (e.g. workstation- or smart phone-based use). Accordingly STUDIO could be easily integrated with any learning management system that should be responsible for user administration and authentication tasks.

2.1   Meta Model of the Domain Ontology

The wide spectrum of ontology applications clearly proves that both the business and scientific world has acknowledged that the detailed exploration of semantic relations must stand at the focus of exploring organizational knowledge—besides the precise definition of concepts (Corcho & Gómez-Pérez 2000; Gómez-Pérez & Corcho 2002). Ontologies modelling domain specific knowledge can also efficiently enhance the integration of information from different sources.

Ontology Language   To effectively support a dynamic conceptual framework, the domain model in the proposed architecture is defined using OWL ontologies (McGuiness & van Harmelen 2004), in which: OWL classes represent such domain concepts that can efficiently support knowledge testing; OWL properties define concept attributes and their relationships; and OWL individuals define concrete domain (such as network management or supply chain management) objects.

Domain Concepts   Our approach mainly foresees the following domain concepts and relations: The Knowledge Area class is at the very heart of ontology, representing major parts of a given domain. Each knowledge area may have several sub-knowledge-areas through the *HasSub-knowledgeArea* inclusion relation. Not only inclusion relations, but order relations connecting knowledge areas in a non-hierarchical way are also important as far as knowledge testing is concerned. In the ontology order is described by the *RequiresKnowledgeOf* relation. For example, if KnowledgeArea$_1$ requires the knowledge of KnowledgeArea$_2$ and KnowledgeArea$_2$ requires the knowledge of KnowledgeArea$_3$, then giving an incorrect answer to any test question related to KnowledgeArea$_3$ is an indication that there is a lack of knowledge concerning both KnowledgeArea$_2$ and KnowledgeArea$_1$.

In order to enable effective knowledge testing, the internal structure of knowledge areas also has to be described in detail. Elements of the Basic concept, Theorem and Example classes form the internal structure of a knowledge area. The *HasPart* inclusion relation connects knowledge areas with their knowledge elements. In order to comprehensively describe the internal structure of the knowledge areas relationship between basic concepts, theorems and examples also have to be identified. The *Premise* and *Conclusion* are order relations that describe basic prerequisites of a theorem (rule or scientific statement) and basic concepts that can be inferred from a theorem. According to CommonKADS[3] methodology such relations have to be presented as option objects in the ontology. These relations are also classes of the ontology that must have special properties, such as the precise description of all attributes on both "ends" of the given relation. While the *RefersTo* reference relation may connect any two individuals of either the Basic Concept or

---

[3] www.commonkads.org.

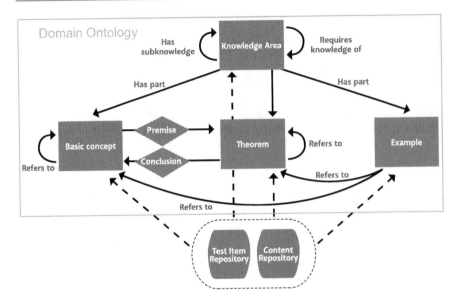

Fig. 2 Meta Model of the Domain Ontology in STUDIO

the Theorem class with each other, individuals from the Example class may also refer to any of the other two knowledge element types. Interrelationships of the major concepts of the domain ontology are shown in Fig. 2, where rectangles represent classes and arrows describe 1-N relations.

The restricted set of relations does not imply limits to knowledge representation but it is a convenient method to improve the computational complexity of the adaptive testing algorithm that has to navigate on the graph provided by the ontology. This knowledge modelling approach is also in accordance with current learning theories. On one hand ontological modelling is nothing other than "connecting specialized information sets and the connections ... that enable us to learn more" (Siemens 2004, p. 5). On the other hand the ontological model of knowledge areas can contribute to the improvement of certain navigating skills of learners—such as creating inferences and analogies, analyzing pieces of information in various ways and making new connections or distinguishing links between fragments of information to create new relations, etc. (Brown 2006).

2.2   Ontology Engineering Components

A series of approaches have been presented in the literature for building ontologies, such as METHONTOLOGY (Gómez-Pérez et al. 1996), On-To-Knowledge (Staab et al. 2001), or Uschold and King's method (Uschold & King 1995; Uschold 1996) to name but a few. According to several methodologies ontology building is an abstraction process where ontology concepts are extracted from an initial knowledge base. Based on other methodologies ontologies are either built from other ontologies, e.g. by automatically generating an ontology skeleton from a huge ontology (Swartout et al. 1997) or by a process of reengineering them.

The first two phases of ontology development—according to the METHONTOLOGY terminology—are: specification and conceptualization. The goal of specification is to determine why the ontology is built and what its intended use is, while in the conceptualization phase the informally perceived view of a domain is converted into a semi-formal specification by identifying the most important concepts of the domain and their relations. These phases are followed by the formalization and maintenance activities. Ontologies built in STUDIO have been used both in education and in business settings for different purposes. The conceptualization phase of building ontologies in STUDIO also varied taking into consideration expectations determined in the specification case.

Building domain ontologies from text is typically used in educational situations in STUDIO where the primary goal of ontology building is to support either knowledge assessment or provide personalized learning experience based on the results of knowledge assessment and the automatic detection of learning styles. In these cases the development of the semi-formal specification is a result of collaboration between the domain expert and the ontology engineer, where curricula, lecture notes and related literature are used as the initial "knowledge base". However, in the case of on the job training, these resources are not available or they do not accurately represent business requirements stemming from corporate processes. Chapter "Ontology Tailoring for Job Role Knowledge" presents a methodology how to extract task specific knowledge from corporate process models and map the extracted concepts into an ontology structure using domain ontologies of STUDIO—if available—as a base. Test mining tools are applied for extracting task related knowledge elements from process models and related documentations.

Building domain ontologies from other ontologies is more typically used in business situations where process ontologies are already available or have to be built to support further applications. Chapter "Corporate Semantic Business Process Management" discusses a semi-automatic, but well-controlled way of enriching domain ontologies using process ontologies. The presented approach describes how to transform the business process into a process ontology and combine it with the knowledge base that is a domain ontology. At the same time Chapter "Future Development: Towards Semantic Compliance Checking" presents how ontology matching tools could be applied in investigating business processes and improving available process ontologies.

Ontology Editor Tools or any other technology enabled tools—in theory—are not required for ontology development, not even in the case of applying the above described meta-model. At the same time the application of ontology editing tools can significantly facilitate the ontology engineering process. These tools were used to build ontologies with ease even without the detailed knowledge or direct application of formalization languages. Moreover, managing a high number of ontology elements, relations, axioms and constraints is also a challenge without adequate computerized aid.

Seidenberg and Rector (2007) have also revealed that there is a need for user-friendly tools able to support collaborative ontology construction and the arrangement of single-users' asynchronous tasks. Usually, there are two major ways in which ontology editing tools support collaborative ontology engineering (Noy 2007). One way is the so called synchronous mode when every user accesses the same version of the ontology and changes are immediately visible to everyone. At the same time in the asynchronous mode users work on their own sandbox space and integrate their changes with the master version later. In some cases editing tools use mixed approaches. Every approach has its own set of advantages and disadvantages.

Besides the above described concerns the *Ontology Editor* in STUDIO also has to meet the following requirements:

- *Extensible*—Due to rapid economic and technological changes business processes and underlying knowledge structures and elements also evolve over time. Accordingly, such a tool is required that would enable the maintenance and development of domain ontologies in an easy but consistent manner.
- *Capable of treating high volume data*—Even one business process or one related curriculum may consist of several hundreds of concepts that must be presented in the ontology. Modelling all the business process-related knowledge elements or all the curricula of a training program will require substantial capacities.
- *Interoperable*—Ontologies may provide a base for different applications, in this way ontology editing tools must be prepared in order to ensure communication and collaboration with other tools in the system.
- User friendly—A simple but consistent interface helps users to work faster and more effectively. Such a tool needs to be developed that besides editing concepts, relations and other properties in a simple way can also provide an easy to understand visualization of the domain ontology. (Szabó 2006).

The STUDIO Ontology Editor follows a mixed approach and provides techniques for both synchronous and asynchronous ontology engineering. A primarily synchronous mode is applied when all the changes made on the ontology are immediately visible for the users as a draft version. Only power users have the right to save the modified ontology as a next version, after checking consistency. There is also an opportunity to use a sandbox space for ontology development.

The editor is also special in the respect that only concepts and relations identified in the above described domain ontology meta-model could be applied in the course of editing. The aim of applying 'built-in' classes and relations was to provide the kind of tool that can be used by domain experts with few or no competencies in ontology engineering. Consequently domain experts can interact with a user-friendly interface where graphical presentation of the ontology also enhances user experience, as shown on Fig. 3.

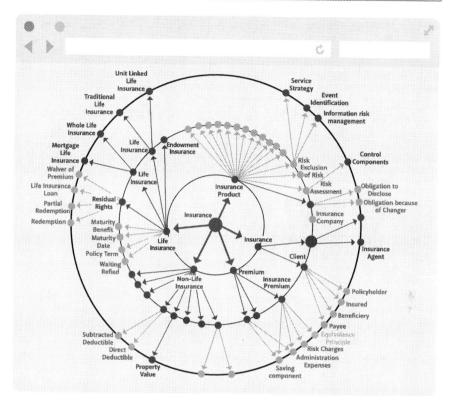

Fig. 3 Ontology visualization in STUDIO—insurance domain ontology

2.3 Content Management Components

The ontology-based domain model is at the core of the STUDIO system as it drives—besides knowledge testing—the creation, storage, query and search for all domain related content as well. Our aim was to provide efficient and flexible content management functionality in STUDIO and to provide a mechanism for developing and maintaining learning materials and test questions in a structured and reusable fashion. The ontology provides the base structure as each single piece of content is connected to one and only one concept of the ontology.

Learning content development starts with the construction of the appropriate domain ontology. As the ontology is finalized, domain experts extend the bare structure with learning materials. Since the structure has already been determined by the domain ontology, the "only" task of the content developer is to assign content elements to the adequate nodes of the ontology. Content elements may have many different formats: images, articles, short texts such as a useful paragraph or a famous quote, audio files or video materials. In order to effectively support learning and knowledge gap fulfillment, learning materials have to be created in such a way that they will also adapt to different learning styles. Visual learners may

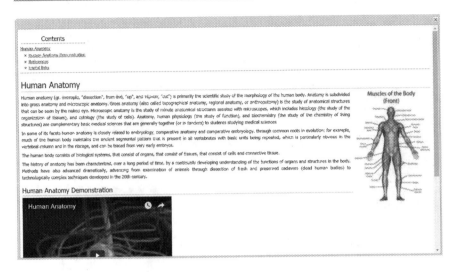

Fig. 4 Learning content in STUDIO

prefer diagrams and presentations, while verbal learners may choose text and lecture notes instead.

The *Content Editor* of STUDIO is a deployment of the Semantic MediaWiki[4] platform that is an extension to the popular MediaWiki engine providing several tools and the special wiki-notation functionality in order to enable the application of ontologies in multiple ways. One of the advantages of using MediaWiki is that it supports multiple types of contents, including text and various multimedia objects broadly used in the learning contents of STUDIO. For Wiki page authors a detailed data formatting and inclusion guideline has also been created, with prewritten html codes. Even if the content developer doesn't have relevant html knowledge it is possible to embed rich media content by simply using a copy-paste mechanism.

The *Content Repository* is responsible for storing and managing these wiki content elements (See an example of wiki content on Fig. 4) and maintaining a rich set of metadata describing them. Each content element can be described with Dublin Core metadata (ISO 2009) and other useful descriptors, like tags or categories. This rich description enables domain experts to easily search the repository for and retrieve already existing contents or create and categorize new elements if needed.

Test item development is crucial in regard to the knowledge testing. In order to adequately support the ontology-based adaptive knowledge testing application every test item must be connected to one and only one concept in the ontology.

---

[4] https://semantic-mediawiki.org/.

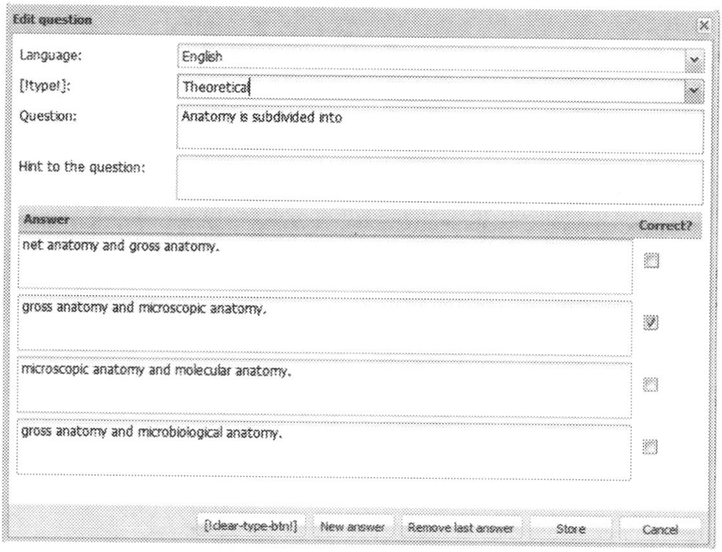

Fig. 5 Test item editor in STUDIO

On the other hand each ontology concept may have several related test questions. In this way the *Test Item Repository* is also structured by the domain ontology. At the same time the Test Item Repository does not form an integral part of the ontology.

Test items are provided in the form of multiple-choice questions. Therefore each test item consists of a question, one correct answer and three false answers. Test item editing and translation into multiple languages is enabled by the Test Item Editor (Fig. 5).

Finally test questions are packaged and deployed in the Adaptive Testing Engine that provides the necessary facilities to execute and evaluate knowledge tests. The ontology is an integral part of the test package, since the execution of tests heavily relies on the underlying ontology structure.

2.4 Packaging Component

The main goal of any application in STUDIO is to create a task or target specific structure extracted from the domain ontology. Accordingly the last phase of content development is packaging, meaning the creation of a set of standard packages that contains the extracted ontology structure of the target (or task-specific) sub-domain, as well as the related learning contents and tests questions. The content package is deployed into the learning management system, while the test package is deployed in the Adaptive Testing Engine.

These target specific structures called *Concept Groups* in STUDIO provide a new layer in the underlying domain ontology. A Concept Group consists of a set of

Fig. 6 Concept group representation logic in STUDIO

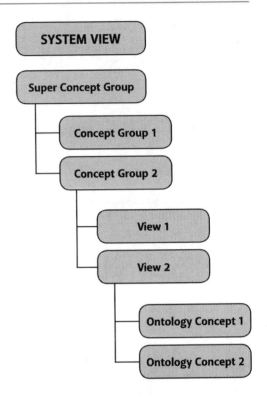

ontology concepts extracted from the domain ontology, and rearranged based on the target of use. The rearrangement does not have any effect on the semantics provided by the domain ontology. The major role of the Concept Group is to enable customization and/or mapping of the domain ontology to the target use case and to provide the basis for adaptive testing.

Figure 6 presents how the Concept Group hierarchy could be built in the STUDIO, where Super Concept Group may represent the organization, a position in the organization or even a training program of an educational institution. The Concept Group represents the target area that should be tested. A Concept group may embody a specific process, a job role or a training specialization. Views provide an entry point to the ontology representing a task or a specific course. Chapter "Ontology Tailoring for Job Role Knowledge" provides an in depth description of the representation logic and possible use cases of Concept Groups in STUDIO.

2.5 Adaptive Testing Engine

Measuring knowledge in a reliable way has always represented a major challenge in training and education. From the 1970s the emerging field of Computerized

Adaptive Testing (CAT) provided important results about adaptive systems that can be combined with semantic technologies (ontologies). In contrast with the traditional examination the number of test items and the order of questions in an adaptive test is only defined in the course of testing with the goal of determining the knowledge level of the test candidate as precisely as possible with as low a number of questions as possible (Linacre, 2000). More precisely, as the test candidates answer the test items, the test "adapts" itself by selecting the next test item to be presented on the basis of performance on preceding items. Adaptive testing is not a new methodology and despite the fact that it has many advantages compared to traditional testing, its application is not widespread. Adaptive tests are usually computer-based tests that have the following main characteristics, independent of the applied testing methodology:

- The test can be taken at a time convenient to the examinee; there is no need for mass or group-administered testing, thus saving on physical space.
- As each test is tailored to an examinee, no two tests need be identical for any two examinees, which minimizes the possibility of copying.
- Questions are presented on a computer screen one at a time.
- Once an examinee keys in and confirms his answer, (s)he is not able to change it.
- The examinee is not allowed to skip questions nor is (s)he allowed to return to a question which (s)he has confirmed his/her answer to previously.
- The examinee must answer the current question in order to proceed onto the next one.
- The selection of each question and the decision to stop the test are dynamically controlled by the answers of the examinee (Thissen & Mislevy 1990).

The current research focuses on the elaboration of such knowledge assessment methodology that enables the exploration of a test candidate's knowledge gaps in order to help them by complementing their training or educational deficiencies. Accordingly, the *Adaptive Test Engine* is a key application in STUDIO that exploits the advantages of ontological descriptions of the domain of interest. As described in Sect. 2.3 every test item resides in the Test Item Repository and is connected to one specific concept in the ontology. In the course of testing the Adaptive Testing Engine "walks through" the ontology structure and asks questions concerning each affected ontology concept. In this way the test candidate's knowledge of a certain set of concepts can be evaluated.

The testing procedure starts the examination at the top of the hierarchy, meaning that those concepts are tested first that have no parent concepts in the given sub-domain (called Concept Group—See Sect. 2.4 for further details). This means that testing typically starts with the evaluation of concepts from the Knowledge Area class. Accordingly, the adaptive test engine provides a testlet[5] related to each top level knowledge area including as many questions that cover the given concept.

---

[5] A testlet is a cluster of test items that share a common path, scenario, or other context.

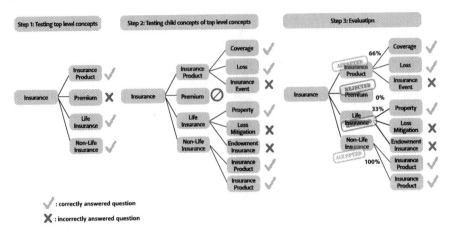

: correctly answered question

: incorrectly answered question

Fig. 7 Illustration of the top-down adaptive testing methodology in STUDIO

For example, if the test candidate was able to correctly answer question (s) concerning the KnowledgeArea₁ in the next stage questions concerning every sub-knowledge area and/or basic concept of the KnowledgeArea₁ will be presented to the examinee. If more than 50 % of all questions[6] (indirectly connected to the KnowledgeArea₁) are incorrectly answered then the KnowledgeArea₁ and its sub-areas will be not accepted. At the same time if certain sub-knowledge area related questions are answered correctly then sub-knowledge areas are tested in the previous manner. In other words the testing engine executes a depth first graph search algorithm in such manner that it closes a branch if the test candidate does not know the given knowledge area or its sub-knowledge-areas and/or given basic concepts at an adequate level. As a result the test candidate's knowledge is thoroughly assessed in respect to the target domain or sub-domain. Figure 7 illustrates the process of adaptive knowledge testing in STUDIO.

Naturally this is not the one and only way in which ontological descriptions could be applied for knowledge assessment. One limitation of the above described adaptive testing methodology is that the test may stop at an early stage, (e.g. in an extreme case, if the KnowledgeArea₁ is the only top level concept in the given Concept Group and the test candidate fails to answer the related question, the test stops and no more questions are presented) which may discourage the test candidate on the one hand while also preventing an insightful exploration of the knowledge structure. For that very reason another knowledge evaluation methodology has also been implemented in STUDIO that follows a bottom-up approach in contrast with the top-down approach of the above described methodology. The bottom-up approach—instead of evaluating single concepts—focuses on analyzing whole assessment paths (that connects a certain concept of the domain with concepts on

---

[6] The examinee can set a threshold according to the objectives of the test to be taken. The selected threshold is automatically applied by the test evaluation algorithm.

the top of the hierarchy). In the first stage of the testing procedure assessment paths are determined, then paths are assessed from the bottom to the top element. If any element in the given path fails, the related ontology concept will be marked as failed that also blocks the current path to the start-element. This failed concept will also block every other path including this element and as such minimizes the set of future sub-paths to assess. The system accepts every path of concepts which reaches the start-element through the relations provided. Incorrectly answered questions and the relating failed concept essentially splits a path into a "top" part which still could reach the start-element and a bottom part which won't be evaluated for the final result.

Chapter "STUDIO: A Solution on Adaptive Testing" provides a thorough review of the theoretical background and organizational relevance of adaptive testing and also presents a detailed description of the adaptive testing methodologies in STUDIO.

2.6   Automatic Learning Style Detection Application

Different people may prefer different ways to learn, or in other words, different learning styles determining how they process and come understand new knowledge. It also has to be taken into consideration that with similar knowledge tested, people may perform differently, depending on the format and focus of the test questions. Truong (2015), through a systematic literature review, indicated that to develop an automatic learning style detection system, a number of stages are required including:

- Learning styles framework selection
- Learning styles attributes selection
- Classification algorithm developments and evaluations

In STUDIO the Felder-Silverman's (1988) learning styles theories have been applied as the framework for the automatic learning style detection application. In the course of STUDIO development, a systematic review was also carried out, which resulted in over 80 potential learning styles predictors. These variables have been being tested, evaluated and engineered. The initial variable selection then becomes the input for the detection model development and evaluation. All of these results, as a consequence, are integrated into STUDIO in the following way: in the first stage, variables from several sources are collected and fed into a data-integration and -processing unit. The output, in the second stage, is used as input for the learning styles detection model, which classifies student' learning styles accordingly. Finally, the information of learning styles of individuals is used as input for a recommendation unit that aids the adaptive functions of the system.

2.7   Content Presentation Component

Feedback plays an essential role in knowledge assessment either by providing advice and recommendation on individual opportunities for improvement or by inspiring motivation. It is also approved in literature that feedback should be personalized rather than using one, general feedback for all. In STUDIO the *Content Presentation* component (See Fig. 1) is responsible for delivering customized materials for end users. Besides presenting the target specific content packages for tutors, domain experts and other power users—created in the Packaging module of STUDIO—that contains the extracted ontology structure, the related test questions and learning contents, the Content Presentation module also provides test results and evaluations, personalized learning content and the statistical analysis of former activities and performance for end users as well.

Learning Material Adaptation plays a key role in enhancing personalized learning experience. The adaptive testing methodology applied in STUDIO enable the repeated identification and fulfilment of knowledge gaps, in order to be able to provide personalised guidance on how the identified knowledge gaps can be effectively eliminated. Monitoring of learning styles are also crucial in developing personalised learning materials and learning activities for end users (test candidates). For visual learners, for instance, diagrams and presentations etc., can be provided, while for verbal learners, texts and lecture notes can be suggested. At the same time this learning style dependent adaptation is still under development in STUDIO since the number of alternative learning materials—supporting every style—still has to be increased.

Evaluation of test results and statistical analysis in STUDIO provides different approaches to follow users' activities and performance. These statistics help end users in making progress towards achieving their learning goals and also help content developers assess the created learning content. Currently, the following functionality is available

- Test evaluation—After each completed test the results are presented and explained in detail, and access is provided to related learning materials.
- User activity analysis—Content developers have the right to analyze how many times test candidates accessed the system, how many tests they have started and how many of them were suspended and/or finished, which questions were included in the test, what the answers to these questions were, etc.
- Data exportation—Content developers can also use the built-in query language of STUDIO to write customized queries for exporting data in a comma-separated value file format (CSV file) for further processing. The query language in STUDIO is based on SQL, accordingly a basic knowledge of SQL is required.
- Connection to external systems—Statistics components of STUDIO could be made available for external systems too. In this case an external system should call this component of STUDIO with an HTTP request, which contains a query. Results can be provided in different formats. In this way results and user

activities can be accessed in the external system as well, without manually exporting data from STUDIO.

STUDIO provides a systematic solution for both controlled knowledge assessment and personalized self-assessment, using a domain ontology to capture the various areas of education providing feedback in multiple ways and in a user-friendly manner.

3 Application Scenarios of STUDIO

In the ontology-centric STUDIO system the behaviors of domain concepts are identified completely using ontological entities, around which different knowledge management tasks could be carried out. Semantic technologies and the underlying applications offered by STUDIO are domain independent, and in this way application scenarios could be elaborated in respect to both business and education. The first scenario was situated in education, while further scenarios were deployed in the Organizational Knowledge Management and Human Resource domains.

3.1 The Educational Setting

Competition in e-learning solutions is increasing at an alarming rate, while social and economic changes and the expectations of both students and the labour market are frequent and diverse. Therefore, there is a great deal of pressure on educational institutions to turn towards the development and application of innovative and modern technologies that enable students to easily access, understand and apply complex curricula and other teaching materials. STUDIO can support education in several ways.

Consensus-based Knowledge Structures are essential in improving interaction among teachers and students. The proposed ontology model (See Sect. 2.1) enables educational institutions to create a comprehensive, unambiguous description of each curricula, or training program of the institution. The resulting domain ontologies are ready to be deployed in managing and improving the educational portfolio and teaching contents of the institution and in enhancing spontaneous learning of students and better understanding of learning materials and their interrelations.

Knowledge Assessment and measuring knowledge in a reliable way is an evergreen issue in education. In order to measure how much students have learned, it is not enough to assess their knowledge at the end of the course. Teachers also have to find out what students know when starting a course. Identifying the prior knowledge of students makes it possible to more precisely identify the knowledge students have gained during the course or training program. The *Adaptive Testing Engine* of STUDIO could be applied both for prior and subsequent knowledge testing. *Concept Groups*—determining the target sub-domain and test package—

can be set up to enable direct evaluation of students' knowledge and performance before and after the given course.

The adaptive testing methodology of STUDIO can also support *self-assessment* providing students with the opportunity to make adjustments to their progress prior to graded evaluation. Taking adaptive tests on their own, students can receive comprehensive feedback on their knowledge gaps. More precisely, a detailed list of those ontology concepts will be provided where the student may have deficiencies according to the test results.

Personalized Learning Experience will only be appropriate if besides supporting what students wish to learn it is also determined how they should learn it. To enable personalized learning, STUDIO system—making use of sthe domain ontology and adaptive test engine—can compile and re-compile self-assessment lessons with personalized sets of learning materials and assessment questions. In the first stage the student's knowledge in respect to the selected domain or sub-domain has to be tested and evaluated in order to identify those (ontology) concepts where the test candidate has deficiencies. Based on the result of this knowledge test a set of personalized learning materials is provided with guidelines on how the learner should "walk through" the ontology structure. In other words, access is provided to the learning material of those ontology concepts where the students incorrectly answered the related test question. Since the results are represented using the ontology visualization tool of STUDIO (See Fig. 3) not only concepts but also their interdependencies are presented to define the proposed paths of learning. The learning experience could be further enhanced—making use of the Automatic Learning Style Detection Application—by adapting learning materials to the learning style of the user.

3.2 The Business Setting

Knowledge acquisition, creation, and transferring together with its sharing have always been a challenge for organizations. It is dangerous to assume that the available knowledge is the right knowledge and it is in the right place. Moreover, the relevance of knowledge may also differ between organizational levels and may also change over time. STUDIO can help organizations to overcome these challenges and to use and reuse organizational knowledge in multiple ways by combining its tools with semantic process modelling techniques (Chapter "Corporate Knowledge Discovery and Organizational Learning: The Role, Importance, and Application of Semantic Business Process Management—The ProKEX Case" provides an overview of the process modelling approach and the related ProKEX solution).

On the Job Training has the benefit of providing direct knowledge and experience for the employee under real working conditions. At the same time this kind of training could be costly since work activities are interrupted by training activities causing delays as well as increasing the number of mistakes. The ProKEX solution—which also makes use of STUDIO's functionality—enables the organization to extract knowledge from organizational processes in order to enrich the organizational knowledge base. This will provide the basis of online, on the job training that allow employees to easily acquire their job role specific knowledge in a customized and efficient manner. More precisely, using the ProKEX toolkit, the domain ontology in STUDIO—representing organizational knowledge—can be improved either by directly extracting business process-related knowledge applying text mining techniques (See Chapter "Ontology Tailoring for Job Role Knowledge") or by matching process ontologies—formally representing knowledge embedded in business processes—with the domain ontology (as also indicated on Fig. 1 and detailed in Chapter "Corporate Semantic Business Process Management").

By using adaptive knowledge tests that are based on the enriched domain ontology the employee's knowledge gaps can be identified, mapped with job role related requirements and addressed with appropriate learning objects. Upon completing an assessment, a knowledge gap report is produced for the test candidate by comparing the knowledge of the employee with organizational requirements. In the event of a discrepancy, the STUDIO system provides the employee with a personalized learning path so that (s)he may improve his/her proficiency level.

Allocating Human Resources is difficult and often fraught with problems despite the fact that there are numerous methods for both short- and long-term resource allocation. At the same time, in most cases implementation issues are not addressed in the literature or the proposed implementation solutions heavily rely on managers' expertise lacking detachment. By using the adaptive testing solution of STUDIO, the knowledge of each worker could be compared with knowledge required by business processes providing an objective basis for matching resource claims with resource offers. As a result, upgraded management of corporate intellectual capital and a better return on investment in human capital can be expected that will lead to more efficient execution of processes and higher improvement in revenues.

Preselection aims at screening suitable applicants where the majority of applicants are eliminated in order to leave only those people most likely to be selected. There are several strategies and tools for preselection (such as Résumés, letters of application, test results etc.) but in any case, job specification and description should form the basis of the applied strategy. Evidently, process models and knowledge extracted from these models provides an objective and complete description of job role related requirements. Accordingly the STUDIO toolkit can provide a knowledge gap analysis of applicants, also enabling mapping test results to current and valid job roles. In the course of preselecting suitable applicants it is important to be unprejudiced and tolerant about the potential each applicant has to be successful in the job. Any specific knowledge gap identified by adaptive tests can be noted and raised during the interview and individually customized learning content can be provided in STUDIO for the applicant if selected for the position.

4   Conclusion

In this work our contribution to organizational knowledge management is three-fold: firstly the proposal of the ontology-centric architecture for developing an extensible knowledge-based system to support the use and reuse of organizational knowledge; secondly the development of a meta-model of the ontology that defines fundamental concepts in a domain independent way; thirdly the development of ontology-based applications to support adaptive knowledge testing, automatic learning style detection, personalized learning both in an educational and business context and human resource allocation and preselection. Further improvements have also been elaborated, designed and prototyped in the context of the ProKEX Project including the application of text mining techniques to enrich domain ontologies (See Chapter "Ontology Tailoring for Job Role Knowledge"), semantic ontology matching (See Chapter "Future Development: Towards Semantic Compliance Checking") and semantic process modelling methods (See Chapter "Corporate Semantic Business Process Management").

Following the completion of several successful pilots, the STUDIO ontology-centric knowledge-based system is being used on a regular basis providing a solid base for maturing the following concepts: (1) Knowledge workers, tutors or teachers cannot be forced to have ontology engineering competencies. Accordingly, a user-friendly ontology editing tool has been developed with a built-in meta-model of ontology. (2) Exploiting the potentials of personalized learning requires the development of alternative knowledge testing methodologies to fit different requirements and the application of learning style detection methods. (3) In order to enable the reuse of organizational knowledge taken into consideration its evolution, as well as knowledge embedded in business process also have to be built into the organizational knowledge base. Therefore, semantic techniques for enriching an organizational knowledge base with process-related knowledge have been developed.

Future works will consist of ontology validation and testing activities in order to improve the application of semantic technologies both in knowledge management and e-learning.

References

Akerkar, R. A., & Sajja, P. S. (2009). *Knowledge-based systems*. Sudbury, MA: Jones & Bartlett.
Brown, T. H. (2006). Beyond constructivism: Navigationism in the knowledge era. *On the Horizon, 14*(3), 108–120.
Corcho, O., Fernández-López, M., & Gómez-Pérez, A. (2003). Methodologies, tools and languages for building ontologies. Where is their meeting point? *Data & Knowledge Engineering, 46*, 41–64.
Corcho, O., Gómez-Pérez, A. (2000). Evaluating knowledge representation and reasoning capabilities of ontology specification languages. In: *Proceedings of the ECAI 2000 Workshop on Applications of Ontologies and Problem-Solving Methods, Berlin, 2000.*

Felder, R. M., & Silverman, L. K. (1988). Learning and teaching styles in engineering education. *Engineering Education, 78*(7), 674–681.

Gómez-Pérez, A., & Corcho, O. (2002). Ontology languages for the Semantic Web. *IEEE Intelligent Systems, 17*(1), 54–60.

Gómez-Pérez, A., Fernández-López, M., De Vicente, A. J. (1996). Towards a method to conceptualize domain ontologies. In: *Proceedings of ECAI-96 Workshop on Ontological Engineering. Budapest, 13 Aug 1996.*

Gruber, T. R. (1993). A translation approach to portable ontology specifications. *Knowledge Acquisition, 5*(2), 199–220.

Hendriks, P. H., & Vriens, D. J. (1999). Knowledge-based systems and knowledge management: Friends or foes? *Information and Management, 35,* 113–125.

ISO (2009). ISO 15836:2009(E): Information and documentation—The Dublin Core metadata element set. https://www.iso.org/obp/ui/#iso:std:iso:15836:ed-2:v1:en. Accessed 10 Jun 2015.

Jarke, M., Neumann, B., Vassiliou, Y., & Wahlster, W. (1989). KBMS requirements of knowledge-based systems. In J. W. Schmidt & C. Thanos (Eds.), *Foundations of knowledge base management* (pp. 381–394). Berlin, Heidelberg: Springer.

Laudon, K. C., & Laudon, J. P. (1997). *Management information systems, organization and technology* (5th ed.). New Jersey, NY: Prentice Hall.

Linacre, J. M. (2000). Computer-adaptive testing: A methodology whose time has come. In S. Chae, U. Kang, E. Jeon, & J. M. Linacre (Eds.), *Development of computerized middle school achievement tests, MESA Research Memorandum No. 69.* Seoul, South Korea: Komesa.

Lucas, P., & Van der Gaag, L. (1991). *Principles of expert systems.* Wokingham, UK: Addison-Wesley.

McGuiness, D., van Harmelen, F. (eds) (2004). *OWL Web ontology language overview, W3C recommendation.* http://www.w3.org/TR/owl-features/. Accessed 21 May 2015.

Noy, N. (2007). What users want: Collaborative development of ontologies. In: *Position Papers of SemGrail 2007 Workshop.* Redmond, 21–22 Jun 2007.

Seidenberg, J., Rector, A. (2007). The state of multi-user ontology engineering. In: *Proceedings of 2nd International Workshop on Modular Ontologies. Whistler, Canada, 28 Oct 2007.*

Siemens, G. (2004). *Connectivsm: A learning theory for the digital age.* http://www.elearnspace.org/Articles/connectivism.htm. Accessed 15 May 2015.

Staab, S., Studer, R., Schnurr, H. P., & Sure, Y. (2001). Knowledge processes and ontologies. *IEEE Intelligent Systems, 16*(1), 26–34.

Swartout, W. R., Patil, R., Knight, K., Russ, T. (1997). Towards distributed use of large-scale ontologies. In: *AAAI-97 Spring Symposium on Ontological Engineering,* Stanford University, Stanford 1997.

Szabó, I. (2006). Implementation of the educational ontology. In: *Proceedings of the The 7th European Conference on Knowledge Management.* Corvinus University of Budapest, Hungary, 4–5 Sept 2006.

Thissen, D., & Mislevy, R. J. (1990). Testing algorithms. In H. Wainer (Ed.), *Computerized adaptive testing: A primer* (pp. 103–135). New Jersey: Lawrence Erlbaum Associates.

Truong, M. H. (2016). Integrating learning styles and adaptive e-learning system: Current developments, problems and opportunities. *Computers in Human Behavior, 55*(Pt. B), 1185–1193.

Uschold, M. (1996). Converting an informal ontology into Ontolingua: Some experiences, ECAI-96 Workshop on Ontological Engineering. Budapest, 13 Aug 1996.

Uschold, M., King, M. (1995). Towards a methodology for building ontologies. In: *Proceedings of IJCAI-95 Workshop on Basic Ontological Issues in Knowledge Sharing, Montreal, Canada, 1996.*

Ontology Tailoring for Job Role Knowledge

Gábor Neusch

1 Introduction

In many organizations corporate intellectual capital is closely related to processes and tasks as manifested in the process models (e.g. in task descriptions) and the related documentations. A task or activity is the smallest part of a process to which human resources can be attached. In other words it can be known who is responsible for its execution. Process models can be augmented with documentations, organizational and other diagrams, policies etc. This additional information can be connected to tasks as task descriptions, so the human capital—knowledge element connection can be made, but unfortunately the documents are usually formalized in an ill-structured way e.g. in simple text format.

In order to compose efficient training courses for employees, or to produce suitable learning materials it is necessary to extract the task specific knowledge from the corporate process models and the attached documentation, and to map it into an ontology structure which represents the domain. During this mapping the knowledge elements of the domain can be restructured based on the process models, and a meta-structure, which is an application oriented contextualization of the ontology, can be created, which represents the relevant knowledge for a specific job role, and in this way can serve as a basis for the training and testing of employees, as a knowledge transfer system.

Knowledge extraction in this case means that those knowledge elements that are needed to perform a specific task properly (task specific concepts), have to be identified. Text mining tools can be used for this purpose with which words and phrases, which represent the task knowledge, can be extracted from the huge amount of ill-structured documentation connected to the process model (Gillani

G. Neusch (✉)
Corvinno Technology Transfer Center, Budapest, Hungary
e-mail: gneusch@corvinno.com

© Springer International Publishing Switzerland 2016
A. Gábor, A. Kő (eds.), *Corporate Knowledge Discovery and Organizational Learning*, Knowledge Management and Organizational Learning 2,
DOI 10.1007/978-3-319-28917-5_5

& Kő, 2014). An underlying ontology can back up the exploration of the related knowledge elements.

The result of the text mining will be a phrase list which can be mapped to a domain specific ontology. **Based on this mapping a process specific representation of the domain ontology can be tailored.** If there are learning materials and questions associated with the nodes of the ontology then this structure can serve as a basis for the automatic creation of electronic tests for a specific job role.

Regarding the elements of the list that results from the text mining the mapping can be successful, i.e. the phrase identifies a knowledge element unequivocally, or is not successful if the given concept cannot be found in the ontology. In the second case it does not mean that the given phrase is a false positive result of the text mining. These elements have to be sent to ontology maintenance as they could be part of the domain conceptual system not yet classified. In this way the domain ontology can be enhanced continuously.

On the other hand, based on the mapping, the process models can be improved continuously as well. Based on the structure and the hierarchy of the ontology the mapping identifies the kind of knowledge elements that were not explicitly formalized in the original process model, but the knowledge of which is very probably needed to perform the tasks properly. In this way the ontology, the employees and the process models can all be improved.

The ontology tailored in the aforementioned way retains the structure of the domain, but in order to use it in a knowledge transfer system (such as e-learning), such meta-structures have to be built based on the outlined subdomains, which allow the flexible connection of ontology parts based on the aim of the usage. In this way a "knowledge map" can be tailored by combining parts of domain ontologies, which can then demonstrate what needs to be known in order to perform a task in an effective and efficient way. An e-learning system based on this kind of meta-structures may be able to discover the knowledge gaps of a person, and this information can serve as a basis for effective training.

The aforementioned method results in a transformation between a process model and a domain ontology. The tasks can be connected to each other in multiple ways. For example, it is possible that the start of one task requires the finishing of another. These dependencies between the tasks may indicate relations between knowledge elements which are connected to these tasks and which cannot be revealed based on the ontology. This is true in respect to the ontology too. Due to its coherence, the ontology may contain several other connections between concepts which are missing from the process models. With this method both the underlying ontology and the process models can be enriched because semantic information can be gained from both which can be used to develop the other. Because of intense automation if there is a change in either one of them it can be easily taken over to the other. In this way their evolution can be facilitated in an intelligent way.

2   The Framework

As shown in the chapter "Corporate Knowledge Discovery and Organizational Learning: The Role, Importance, and Application of Semantic Business Process Management—The ProKEX Case", the goal of the ProKEX project is to facilitate the knowledge transfer in a company by eliciting the necessary knowledge elements from the task descriptions and the documents related to the process models, mapping this knowledge in the form of words and phrases to a domain ontology, and tailoring task specific meta-structures that are used as a 'learning path' to train employees. In order to achieve this goal several tools were used to manage the life cycle of the knowledge management.

The most important technical components of the suite are the process modelling tool, the text-mining application, and the ontology learning tool which is basically an interface of the domain ontology located in the STUDIO system. The STUDIO system is an ontology management and adaptive e-learning system. Besides the domain ontologies, STUDIO stores the Concept Groups. A Concept Group is a new layer upon the underlying ontology, a tailored meta-structure which follows the logic of the application coming from the processes. For more information about STUDIO see chapter "STUDIO: Ontology-Centric Knowledge-Based System".

Process modelling is explained in detail in chapter "Corporate Semantic Business Process Management", in Gábor et al. (2013) and in Ternai et al. (2014). Chapter "ProMine: A Text Mining Solution for Concept Extraction and Filtering" gives a detailed description of the text mining component used in ProKEX.

The result of the text mining is a list of words and phrases (task specific concepts) which comes from two sources. Either a phrase comes from the original source of information—specifically from the process models or the related documentation—or it is the result of the enhancement of the corpus which was carried out during the text mining. This list can be extensive due to the high number of external sources that can be used during the text mining process, and can contain a number of unrelated terms in spite of the dimension reduction methods. In order to use the result of the text mining in a knowledge transfer system it is necessary to remove those unrelated phrases which have no meaning or are not related to the fields of job knowledge under examination. To achieve this the next stage of the suite is the mapping of these results and the domain ontology. The mapping stage takes place in the STUDIO system.

During the mapping, those concepts (ontology nodes, individuals of the STUDIO ontology) are selected from the ontology, which are relevant for the tasks of the process models. If the model of the knowledge which needs to be known in order for the tasks to be created, the structure of the necessary knowledge for a process or a job role can be built. "Relevant concepts" are identified by elements of the list of phrases resulting from the text mining. The phrases, which are grouped according to tasks, may, from a logical point of view, represent knowledge elements which are needed during the given task (task related). On the other hand from the technical point of view, a phrase can identify a node (ontology concept) in the STUDIO domain ontology. The mapping process discovers one-to-one correspondences between the knowledge elements of tasks and ontology concepts.

The result of identification can be a total or partial match and non-identified as a type. "Non-identified" is a knowledge element (a phrase from the list achieved by the text mining) if there is no ontology concept which correspond with that given knowledge element. As the phrases originated from documents written in a natural language, then, despite the preprocessing, several exceptional cases can occur because of the weaknesses of the language processing or due to human error. These include, but are not limited to:

- The input of the text mining and the ontology elements are defined in different languages.
- The terminology or the qualifiers are used differently.
- Synonyms or periphrasis are used.
- Suffixes or prefixes are used differently.
- Misspelling or typos.

Another question is how the identity of a phrase and an ontology element is defined. Within the framework of the ProKEX suite a textual comparison is made between a phrase and the description of the ontology nodes, the description of which is basically an appellation of the given concept. Based on the aim of the application several strategies can be imagined. One of the possible strategies is that only the full matching can be accepted irrespective of case sensitivity, after special characters were removed. The other possible strategy is when partial matching is accepted as well, but in this case several new questions occur such as how it is defined based on part of the phrase or a syllable, or simply on a percentage ratio. So during the mapping the phrases extracted by the text mining identifies fully or partially the ontology nodes. Within the framework of the ProKEX project the fully identified ontology concepts are used mandatorily in additional processing, and the system lets the knowledge engineer select which partially identified nodes (s)he wishes to use in the succeeding stages. It is also possible that no ontology node was identified in the STUDIO domain ontology by a phrase of the list achieved by the text mining. These phrases are forwarded to the ontology maintenance, where those which have relevance in connection with the domain may be classified into the ontology since they can carry relevant information and therefore may have an added value (Fig. 1, Point 6.).

For the topic, which was appointed by the ontology concepts identified in the succeeding stage of the ProKEX process, additional nodes are sought in the ontology that may be relevant for the topic, in order to enrich the set of nodes, and in this way make the topic "sharper". This enrichment is based on the relations of the ontology, and the distance between the ontology concepts. The enriched set of ontology concepts are composed into a Concept Group. The aim of this stage in respect to human resource management is to build the system of knowledge that needs to be known in connection with a given position (job knowledge). This is why the task related knowledge elements are identified in the ontology and restructured

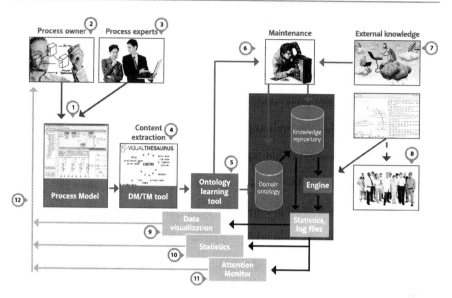

Fig. 1 The overview of the ProKEX Project

into a position specific view. Creating Concept Groups has a practical reason. The structure and the hierarchy of the ontology are organized based on the logic of a domain. For the application of the semantic strength of the ontology, a control structure is necessary which makes it possible to reorganize the ontology concepts into an applied point of view. In the STUDIO the hierarchical structure of the Concept Groups allows the trainer to customize the ontology, and decide which ontology concepts need to be asked during the test, as there can be questions and learning materials attached to these nodes in the aforementioned way. By using the Concept Group the knowledge engineer is able to tailor together several sub-domain ontology parts, and it is then possible to reorganize these sets of knowledge elements in a flexible manner.

The Concept Group system is the basis of adaptive testing. The STUDIO requests an answer for the questions connected to the tailored ontology concepts based on the hierarchy of the Concept Group during a test. The aim of the test is to discover the "black spots" of the test candidate, in other words to point out those knowledge elements which are not known by the user but which would be needed to perform in a job role properly. It can be seen whether an employee has the necessary knowledge to perform his or her roles, and if not, by using the STUDIO the missing knowledge elements will be highlighted, and the missing parts can be made up easily. Testing, i.e. the way the knowledge of an ontology concept is checked, could be done in several ways, based on a diverse range of algorithms. In STUDIO testing goes from the more general knowledge areas to the specific ones by default, but research has been carried out on working out new testing methods and algorithms. To find out more about the testing methods see chapter "STUDIO: A Solution on Adaptive Testing".

3   Technical Overview of the Concept Group Generation

As outlined previously, in the STUDIO system the basis of the adaptive testing is the Concept Group, which is the control structure for the testing. It is a technical structure, with which the ontology can be tailored and the trainer can define which part of the "universe of the ontology" should be requested from the trainee. The Concept Group can therefore be imagined as a simple hierarchical regrouping of the ontology, which defines a taxonomy based on the application (testing). The Concept Group is a very flexible object of the system and it provides plenty of opportunities to those who compose an adaptive test. The Concept Groups can be grouped in several levels and it offers a complex modelling framework for the organizational knowledge. A Concept Group represents the knowledge of an organizational process, or it models job knowledge or even a major at an educational institution as shown in the Fig. 5.

A Concept Group is connected to the ontology with the 'Start Node' object, which in a technical respect is an interface toward the structure of the domain. In the STUDIO there are two technical classes created for this purpose the 'Curriculum' and 'View' classes.

Although STUDIO ontology design follows the 'Closed World Assumption' principle, it is also possible to define new concepts in the system. One of the functions of the Curriculum class is to serve as a definition point for a sub-domain ontology. The reason why such a definition point is needed is because the connection between distinct parts of the ontology may not yet have been defined. A knowledge element which refers to a specific node in the system can easily be brought into being through their connection, and it will be classified into sub-domains. In other words, if the knowledge engineer wants to define a new ontology concept the node has to be found into which the concept fits. Next, the concept will be classified through a hierarchical or a symmetric relation. An example of the hierarchical relation is the 'Has sub-knowledge area' relation, whereas for the ulterior one it is the 'Requires knowledge of' relation. Technically speaking a concept can only be created through a connection. There cannot be an ontology element in the STUDIO that stands by itself. In order to create a new concept which has no related nodes in the existing ontology, a new topic has to be defined which represents a new sub-domain.

In order for a knowledge engineer to be enabled to define new sub-domain ontologies, a technical class—a definition point for the ontology—is needed. This class is the Curriculum. If the knowledge engineer connects a Curriculum into a Concept Group as a Start Node, it will create an entry point in the sub-domain ontology which is represented by the given Curriculum; and allows the tailoring of the ontology through its structure with the hierarchy of its nodes retained. Therefore the Concept Group composed in this way has to very strictly follow the logic of the underlying ontology.

Sub-ontologies can intersect each other. To illustrate this let us assume that the ontology contains two sub-domains, e.g. 'Environment Protection' and 'Informatics'. The two sub-domains may be connected e.g. through the 'Control Software of

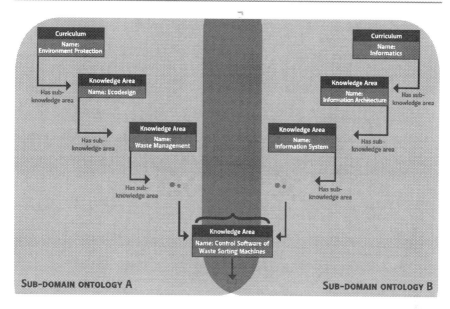

Fig. 2 Intersecting sub-domain ontologies

Waste Sorting Machines' Knowledge Area (Fig. 2). For more information on the Curriculum class see chapter "STUDIO: Ontology-Centric Knowledge-Based System".

Retaining of the structure means that between the elements of the Concept Group and a Curriculum as a Start Node of the Concept Group there has to be a route in the ontology. In other words, in respect to the testing, this means that if the knowledge engineer wants to ask a specific knowledge element from the test candidate, then all the more general knowledge elements have to be asked, which are on the path through the Curriculum because it has the role of a sub-domain ontology definition point and a role of Start Node (entry point for a Concept Group) in one object.

In order to understand the topic of the hierarchy—and how it works in connection with the Concept Group system—it is sufficient to imagine the STUDIO ontology as a directed rooted tree, where the roots are the Curriculums, the sub-domain definition points, and the directions point from the more general knowledge elements to the more specific ones (Fig. 2). If sub-domain ontologies are connected to each other through an ontology concept it is not allowed to select a node which represents a more general knowledge area in the inverse direction (against the allowed direction), during the Concept Group tailoring (Fig. 3).

It would contravene the convention of the adaptive testing, namely that the testing always goes from the more general knowledge areas to the more specific ones, which are only asked if the question of the parent node was answered correctly, i.e. if the general idea is well known by the test candidate. For example if we connected to a Concept Group the Curriculum of the 'Environment

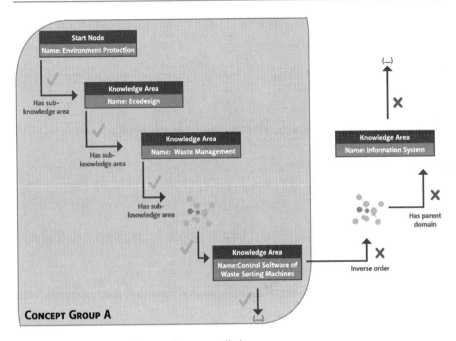

Fig. 3 Restriction of the Concept Group compilation

Protection', and from this Start Node we selected the knowledge area of the 'Control Software of Waste Sorting Machines', we would not be able to select additional elements—in this respect upwards in the hierarchy—to the direction of the Curriculum of the 'Informatics' sub-domain, even if both Curriculum objects are connected to this knowledge element (Fig. 3). In this case if we also want to ask the elements of the other sub-ontology of 'informatics', then this instance of Curriculum has to be connected to the Concept Group as a Start Node as well. In this tailoring conception it is possible that there are elements connected redundantly into a Concept Group, and of course such a situation must be managed in the follow-up processing as exceptions.

One of the added values of the ontology comes from the hierarchy of its elements, so there are no circumstances which would give reason to ease the restrictions given by the rules outlined above. But in connection with the structure there can be such circumstances. It is possible, for example, that the editor of the test would like to connect some specific knowledge element to a Concept Group, but the routes through which those specific nodes can be approached are irrelevant in regard to a specific task or job role for which the test is composed. In order to make it possible to omit irrelevant nodes from the tests in respect to the application another technical class was defined; i.e. the 'View' in the STUDIO system.

Practically the View is a technical object underneath which knowledge elements can be gathered. It acts as a Start Node, i.e. an "entry to an arbitrary point in the ontology", and in this way ensures the structural freedom in the design of a Concept

Group, but after it is connected to the ontology it retains the added value of the hierarchy.

Logically speaking neither the Curriculum nor the View is an element of the ontology, these objects are merely interfaces through the use of which knowledge elements can be connected to a Concept Group. The difference between the two—what restricts the scope of the accessible knowledge elements from a Curriculum—comes from the technical realization. Although from a logical point of view the Curriculum is not a class of the ontology, it still is technically, because if a completely new domain needs to be defined in the STUDIO ontology, which is not connected to any of the existing parts, then it can only be done by using this technical element. Thus, in practice if we wish to define a knowledge element in the ontology that cannot be connected to any of the existing nodes in a logical way, it can only and exclusively be connected to an instance of the Curriculum class. This is because an ontology concept can only be defined thorough a relation in the system. In this way the Curriculum object has two roles in the system. As a sub-ontology definition point, it allows the definition of new domains, hence the knowledge elements connected to it have to follow the structure dictated by the logic of the domain. As a Concept Group Start Node it makes it possible to tailor the connected subdomain ontology based on the logic of the application, but the structure given by the domain cannot be by-passed. On the other hand View is not part of the ontology either logically or technically. It is not allowed to define a new knowledge element from a View object since its only role is to serve as an interface from the Concept Groups to the ontology.

4 Representing Concept Groups

The STUDIO ontology is stored in a Tokyo Cabinet (FAL Labs 2010) key-value pair database. In the database every Concept Group object is represented by several key-value pair records, the most important of which contains the elements of the given Concept Group. This record contains a structured, JSON text object the logic of which can be seen in Fig. 4.

In this structure, every element of the outermost array object represents a tailored sub-ontology part which is connected to the Concept Group through a Curriculum or a View as a Start Node. The value of the '*members*' array in each element of the main, peripheral array consists of the identifiers of the ontology nodes tailored to the Concept Group as a set, while the value of the 'ownerNode' is a pointer to the Start Node which is the interface to the sub-ontology (Curriculum or View).

As can be seen in this example, no information about the ontology hierarchy is stored in the STUDIO database in connection with a Concept Group. The only necessary information to store here is a pointer which shows the entry point to the ontology. After this point every information about the hierarchy is given by the

```
1   [{
2          "members": ["StartNode1",
3                      "KnowledgeArea3",
4                      "KnowledgeArea1",
5                      "KnowledgeArea2",
6          "ownerNode": "StartNode1"
7   },
8   {
9          "members": ["StartNode2",
10                     "KnowledgeArea3",
11                     "KnowledgeArea4",
12                     "KnowledgeArea5"],
13         "ownerNode": "StartNode2"
14  }]
```

ontology itself, and in connection with the Concept Group only the elements of the meta-structure, i.e. the IDs of the tailored nodes have to be stored. For the sake of understanding the technical representation it has to be noted that every instance of every ontology class and also the Views have records for storing their neighbors through every relation. These are lists of nodes which store the horizontal and vertical (hierarchical) links of the ontology concepts. For more information about the defined relations of the model of the STUDIO ontology see chapter "STUDIO: Ontology-Centric Knowledge-Based System".

5   The Modelling Logic

As can be seen from the example data structure in the previous section, more than one Start Node object can be connected into a Concept Group and they do not even have to be in the same class. In this way composers of tests are given a highly flexible toolkit to customize their tests. In the STUDIO system Concept Groups can be grouped further into a super class dubbed the 'Super Concept Group'. In this way the hierarchy of the Concept Groups can model the real world in respect to Human Resource Management (HR), education or to a process point, as shown in Fig. 5.

From the Human Resource Management point of view, every employee has a position, in which several job roles have to be fulfilled. In order for the employee to be able to execute his or her job roles properly, they need several competencies that can be well defined by using the knowledge elements. So it can be said that a person has a given competency if and only if (s)he has a certain set of knowledge.

In formal education majors may consist of one or more minors. From this approach it can be said that in order for a student to be able to pass his/her exams in connection with a specific minor he must have several competencies, so just like

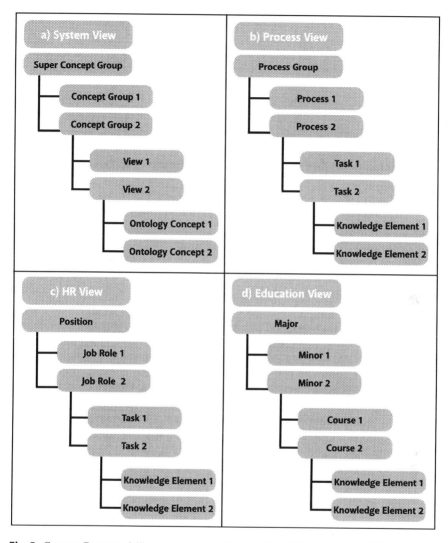

Fig. 5 Concept Group modelling overview—(**a**) System view, (**b**) Process view, (**c**) HR view, (**d**) Education view

from the Human Resource Management point of view he needs a well-defined set of knowledge.

Corporate knowledge can also be modelled from a process point of view if it is assumed that the organization has to execute some well-known and definable processes in order to achieve its mission. Human resources can be allocated to tasks in the processes. As tasks are well-defined stages of activities, in this way tasks also contain descriptions about what to do and how to do it. So it can be said that the person who is responsible for the execution of the given task must be familiar with certain knowledge elements in order to do the task properly. As mentioned previously, in the STUDIO system the basis of the adaptive testing is the Concept Group. Hence, if we are able to build a Concept Group system representing this process view from the result of text mining of the documents connected to the tasks of the processes, then we will be able to make evaluations based on the test results, whether an employee has the knowledge necessary for fitting in with the job roles connected to a given task or not. It will also be clear which knowledge elements are missing, in other words where the gaps are in the employee's knowledge. In order to be able to regroup the knowledge elements into a process view the task must be retained as an attribute during the preceding stages of the procedure.

It is possible that in order to execute certain tasks the cooperation of several employee is needed who can act in one or more job roles, and such employees would need the knowledge of several well defined knowledge elements to fulfill their tasks properly in the process. One employee may perform several tasks in connection of his or her position, and in a process may appear in several job roles. These job roles also can be connected with the tasks in the process models. If we are able to retain this connection as an attribute during the text mining then the coherence between the job roles and the knowledge elements can be discovered, and the fitting Concept Group can be built from HR point of view based on the job knowledge instead of competencies. If a proper organizational model was also connected to the process models then it can be ascertained which job roles are connected to which position. In this way, after all of the organizational process models have been run through the ProKEX suite the Concept Groups representing a position can also be built. **This is the target from a human resource management point of view, i.e. that the knowledge area structure can be built based on a position, since in most organizations, employees are hired for a position.** If all the knowledge needed for the job roles of a position can be asked in the system, and the knowledge of the employees can be tested in this level of aggregation, it can facilitate the selection processes of the organizations, in order to find the employee who is the most suitable for a given position. The system can be used for internal training purposes as well, as the tests based on Concept Groups built in the aforementioned way are able to discover the knowledge gaps of the employees, hence they can complement their deficiency.

In order to achieve this target it is necessary for the models which represent the organizational processes to be ready in an appropriate level of detail, with the other connected documents, such as the organizational diagram. These documents and models need to be connected in a task level. In other words the process models need to be properly annotated.

The text mining results of the process models are such lists of potential knowledge elements, where the list items are connected with the task and the job role—which need to be known—as an attribute. A knowledge element may be needed for several tasks, or job roles. A possible example of the result of the text mining can be seen on the Fig. 6.

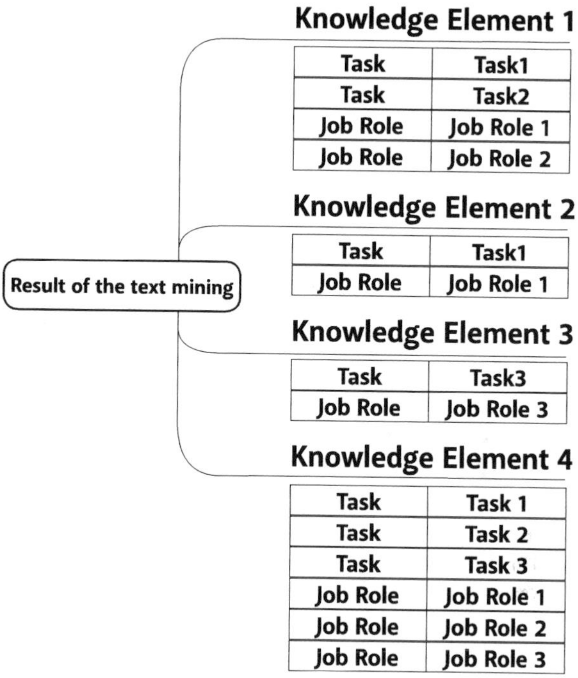

Fig. 6 A possible output of
the text mining

Knowledge Element 1

Task	Task1
Task	Task2
Job Role	Job Role 1
Job Role	Job Role 2

Knowledge Element 2

Result of the text mining

Task	Task1
Job Role	Job Role 1

Knowledge Element 3

Task	Task3
Job Role	Job Role 3

Knowledge Element 4

Task	Task 1
Task	Task 2
Task	Task 3
Job Role	Job Role 1
Job Role	Job Role 2
Job Role	Job Role 3

A Concept Group can be built from this result based on several modelling points of view to support the application logic. Should the retrieved knowledge elements have to be depicted based on the positions, in other words, if the aim of the test is to assess the knowledge of an employee in connection with a specific position, the following stages need to be followed in STUDIO.

1. As can be seen in Fig. 6 there must be defined View for every task in which the employee works in connection with one of his/her job roles.
2. The ontology concepts tailored on the basis of the result of the text mining have to be connected to the View created in this way.
3. The Concept Groups representing the job role (this will be the basis of the test) and the Super Concept Group exemplifying the position also have to be created.
4. As a last stage the Concept Groups have to be tailored on the basis of the hierarchy given by the ontology spring from the nodes connected to the View on the first level.

In this way a degree of simplification was achieved. It is assumed that somebody is able to perform a job role properly if all the knowledge elements needed for the tasks connected to the specific job role are well known. Tasks in process models are annotated with job roles, thus it can be ascertained who is responsible for the execution (cp. RACI). Due to the aforementioned, and based on the process models, redundant lists of tasks grouped according to job roles can be defined. As can be seen in Fig. 6 in this case tasks are represented by Views and the first level of ontology nodes can be connected to these objects as ensued from the aforementioned assumption.

6   Structure and Hierarchy

During the building of the Concept Group system the ontology tailoring STUDIO uses the ontology concepts which were identified by the phrases which were derived from the text mining. In connection with these concepts several attributes are known, such as job roles and tasks, as depicted in Fig. 6. If nothing more is done, only just the connection between the identified ontology concepts and the Views representing the tasks is made, and STUDIO would be able to ask them as a test. **Within the framework of the ProKEX project the term 'mapping' is used to name the process, when the phrases are mapped to the ontology. During the mapping the connections between the identified ontology concepts are discovered, and in this way the set of phrases which describe the necessary knowledge for the tasks are enriched with semantics. Other ontology concepts which may be connected to the given topic based on the structure of the ontology may be attached as well.** Several strategies can be envisioned for carrying out the mapping.

1. Retain the whole structure and hierarchy of the domains which were pointed out by the ontology concepts identified through text mining.
2. Retain the hierarchy between the ontology concepts which was found, but omit the structure given by the domain.
3. Discover possible routes between ontology concepts based on a graph distance, the extent of which was set by a knowledge engineer.

Retaining the structure (point 1.) means that the Concept Group is tailored on the basis of the logic of the Curriculum concept and the adaptive testing. In this case there always has to be a route between a given node selected for the Concept Group, and the Curriculum element represents the entry point toward a sub-domain ontology. In this way a lot of ontology concepts may be selected automatically that represent more general knowledge and which do not necessarily need to be known to execute a task properly. Based on the logic of the adaptive testing used in STUDIO, if these more general knowledge elements are answered incorrectly it is possible that those elements which were explicitly pointed out during the text mining were not even asked.

Should the structure be retained Curriculum elements could also be used as Start Nodes, as this object has the ability to interface a sub-ontology into a Concept Group while the structure and the hierarchy are retained. But in practice the View object must be used because tasks are defined in the Start Node level, and in order to model a task comprehensively knowledge elements could be needed from several sub-ontologies. An example of this kind of tailoring logic is shown in Fig. 7 and the Concept Group created based on shown in Fig. 8.

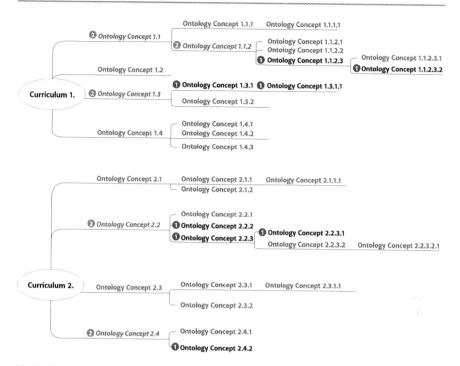

Fig. 7 Example sub-ontologies I

The Ontology concepts which are in bold and marked with '1' in Fig. 7 are representing the nodes which were explicitly pointed out by the output of the text mining in connection with a hypothetical task. If the aforementioned logic is followed the nodes which are brought into prominence with italic font style and marked with '2' are also automatically selected from the ontology. The sub-trees selected by this algorithm can be connected to the View object which represents the given task as shown in the Fig. 8, where the View object is in bold and underlined. The system view of Fig. 8 is shown in Fig. 5.

A Concept Group tailored on the basis of the aforementioned algorithm retains the structure and the hierarchy of the ontology. In other words, in addition to the structure of the domain built in the STUDIO ontology being preserved, the logic of the adaptive testing is also followed. Using this logic, however, poses some challenges, and raises questions in respect to logic as well as certain technical points.

In respect to logic, it is a legitimate question whether during the test based on a Concept Group created in this way it is really necessary to ask questions about every more general knowledge area as well in order to ascertain whether the knowledge of the test candidate is enough to perform the task? Do the employees really need to know the whole broader domain, or can they perform a specific task properly without it? A definite answer cannot always be given for this question. It depends on the position and the domain as well. From a technical point of view the

Fig. 8 The Concept Group representing the results of the mapping based on mapping strategy 1

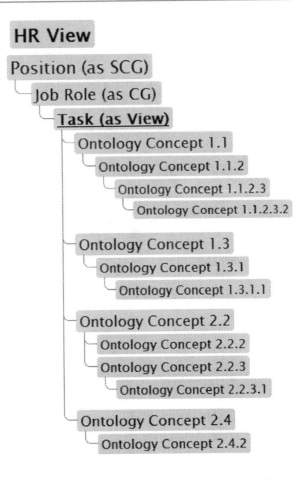

kind of solution can be suggested in which the tailoring strategy can be chosen based on the aim of the usage.

From a practical point of view, determining the relevant Curriculum for a given node is a challenge, if it is an element of more than one sub-ontologies such as in the previous example; the 'Control Software of Waste Sorting Machines'. The question in this case is the broader knowledge of which sub-domains' are relevant for the given application? During the mapping process it can be checked what knowledge elements of which domain dominate the results of the text mining, or the inverse relations of the STUDIO can be used, but it is really hard to cope with this issue, and the possibility of an exceptional case is high.

In the case of point 2, where the hierarchy is retained and the structure is omitted, the super- and subordinate relationships would be retained between the ontology concepts, but it is not necessary to have a route between a specific node and the Curriculum object (sub-ontology definition point). In this case the more general knowledge elements which were not explicitly found during the preceding processing are considered as irrelevant and the structure dictated by the domain

Fig. 9 The Concept Group representing the results of the mapping based on mapping strategy 2

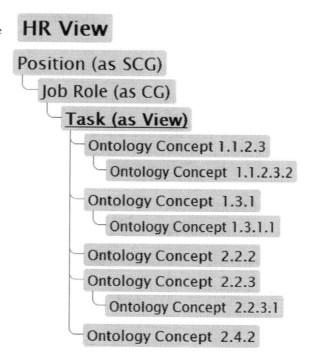

ontology is omitted from the Concept Group. In this case the elements and their relations have to be discovered in the ontology. If there is a connection between two elements, sub-trees can be created based on the hierarchy which was stored in the ontology. In this case only those elements would be part of the Concept Group which were explicitly revealed by the text mining process. If the example ontology of Fig. 8 is considered as a basis, the object in this case is to find the relations between the nodes marked with bold and '1', and create the Concept Group as shown in Fig. 9. In this way only those questions are asked during the test, which are connected to an ontology concept that was explicitly identified in the task description or in the connected documentation.

This tailoring strategy only retains the relevant concepts and the relations between them but do not use the full semantic strength given by the ontology.

A third tailoring strategy could be if discontinuities of the ontology—the distance of which is set by a knowledge engineer—are spanned together along its structure in spite of the fact that the mid-elements were not identified in the original documents like process models etc. An example of this strategy is shown in Fig. 10.

The elements in bold and marked with '1' represent the ontology concepts that were identified on the basis of the phrases extracted by the text mining. If the possible distance of discontinuity is set to one element, which is two if we consider it as a graph distance, then the elements made in italics and marked with '2' are also selected from the ontology and connected to the Concept Group system, even if

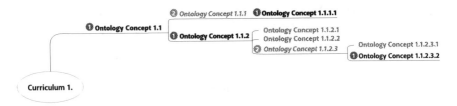

Fig. 10 Example sub-ontologies II

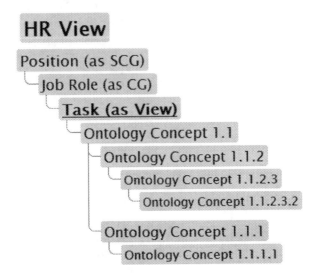

Fig. 11 The Concept Group representing the results of the mapping based on mapping strategy 3

those nodes were not explicitly pointed out by the text mining. In this way the Concept Group hierarchy showed in Fig. 11 is tailored in the STUDIO.

The extent of discontinuities, in other word gaps, which can be spanned, must be adjustable. **Spanning the gap within a 'reasonable distance' can carry semantic added value.** In this way those ontology concepts can be discovered and connected to a Concept Group as well which were not found explicitly in the process models as knowledge elements, but based on the structure of the domain ontology they are elements of the well-defined topic, which is outlined by the nodes which were identified based on the result of the text mining. **Additionally, if these mid-elements can be sent back to the process designer, the process model itself (e.g. the task descriptions) can also be improved, enriched.**

The idea behind this concept is based on a well-defined 'topic' (in the ontology) being hedged by two ontology elements a knowledge of which are unequivocally needed to carry out a given task. The beginning and the end of this 'topic' in this way are explicitly marked out by the process model or the documentation. It is logical and reasonable to think that it is also important to know the mid-elements in order to execute the task properly. It is reasonable to assume that if an employee needs to know the knowledge behind 'Ontology concept 1.1.2' and 'Ontology concept 1.1.2.3.2' (Fig. 9) in order to properly perform in a task, it is highly probable that the knowledge of the bridging 'Ontology concept 1.1.2.3' would also be needed, at least at the level of comprehension. Technically speaking, based on a distance value set by the knowledge engineer, possible routes are calculated between the identified nodes of the STUDIO ontology. This measure is based on graph distance, which is calculated between two extracted concepts found in the ontology. If two concepts are within the distance relative to each other, we assume that the nodes between them are also relevant for the context.

It is really hard to determine the highest discontinuity between two ontology concepts which can be spanned together on logical grounds. In other words how the extent of the relevant topic can be determined. It would be hard to automatize this decision because the distance of two nodes are based on the interaction of several things like the task itself or the domain ontology. In other words there may be some cases when the mid-elements tailored by the topic are relevant, and have a huge added value, but in other cases these nodes may be completely irrelevant. So it is impossible to set a universal distance based on which meaningful tailoring will always be carried out. The ideal distance should be spanned, i.e. the borders of the relevant topic always change based on the application, and have to be set by a knowledge engineer, based on expert judgment. Based on experimental data in the ProKEX project the distance which was used is three, because it was fit for the purpose almost every time.

In the ProKEX project these three tailoring strategies were considered, and based on the aforementioned pros and cons an algorithm for the third strategy was developed. The logic of this strategy provides a great deal of flexibility when a Concept Group is tailored based on the result of the text mining, but retain the added value given by the hierarchy between the nodes built in the STUDIO ontology. It may also serve as a basis of improvement of both the process models and the ontology.

7   The Algorithm

The aim of the algorithm is to tailor the aforementioned 'topics', to ascertain which element of each topic is 'ranked highest' in the hierarchy of the STUDIO domain ontology (technically which element is the closest to a sub domain definition point) and to connect these elements to the View object which will represent the task.

1. The algorithm loops through the list of ontology concepts (*list 1*) identified by the phrases which are mined from the process models.
2. From each node (*onode 1*) of *list 1* the ontology is traversed through recursively along the inverse relations (upward hierarchy) until the adjusted distance become zero, which is decreased in each stage. The ontology concepts, which

are touched on the path, are stored in a temporary set object, which can contain an element only one time. *Onode 1* is removed from *list 1*, what is important because of the recursive nature of the algorithm.

- If another element (*onode 2*) *of list 1* is found on the path passed through the ontology, both *onode 1* and the elements of the temporary list are added to a set which contains the elements that will be added to the Concept Group members' array (*list 2*). The elements of the temporary list are added to another set which only contains the mid-elements (*list 3*). This list will be needed later. The temporary list is cleared. Point 2 starts again but now the starting point will be *onode 2*.

- If the distance was decreased to zero and each possible direction was inspected in the ontology and *onode 1* is not a member of *list 2*, then *onode 1* is added to a list which contains all the nodes which have to be connected directly to the View object (*list 4*), *onode 1* will be added to *list 2* and the algorithm stages further to the next element of *list 1* (point 1.). If *onode 1* is a member of *list 2*, then the algorithm just stages further to the next element of *list 1* (point 1.), without putting *onode 1* into *list 4*.

3. The algorithm iterates through the list of the mid-elements (*list 3*). From every element of *list 3* the ontology is walked through the hierarchy downwards until the adjusted distance becomes zero, which is decreased in each stage. The elements on the path are stored into a temporary object, which is cleared after the distance reaches zero. If an element (*onode 3*) of the set, which contains all the nodes that have to be connected directly to the View object (*list 4*), is found on the path passed through the ontology, *onode 3* is removed from *list 4*, and the elements of the temporary list are added to the list which contains the elements that will be added to the Concept Group members' array (*list 2*).

In this way two sets will be available when the algorithm finishes. The first will contain all the nodes that have to be put to the members' array of the Concept Group (*list 2*). It is not necessary to store the relations between the nodes because this information is stored and available from the ontology. The second set will contain the nodes that have to be connected to the Start Node (*list 4*). As a final stage the technical objects (View, Concept Group) have to be created, and the necessary connections have to be made in the database.

The outlined algorithm explores the ontology until the extent that the distance measure was set, and the topics enclosed by ontology concepts are discovered. Hence the semantic value of the ontology is utilized. Based on the Concept Groups tailored for a position, knowledge tests can be provided for employees, and the knowledge gaps in their job knowledge can established. The mid-elements discovered by the algorithm can be sent back to the process experts. Based on the learning materials in respect to the STUDIO ontology concepts, the task description could also be expanded as a 'side effect' of the ontology tailoring.

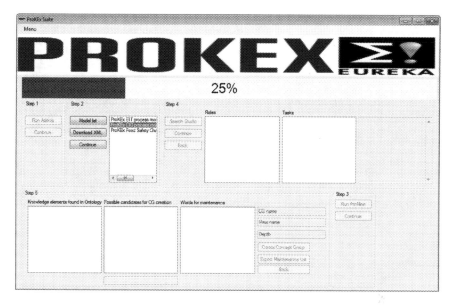

Fig. 12 Insurance processes are selected from the process list

8   Realization of the ProKEX Suite

In this sub-chapter the beta version of the ProKEX suite will be presented through an example application from the Insurance domain. This example was taken during the pilot testing of the project. The goal of this chapter is to present the outlined logic and the way in which the algorithm works in connection with a real world problem.

The whole 'ProKEX process' is controlled with a desktop application which integrates all of the components of the suite. The process modelling is the first stage of the work which is done in the external Adonis BPM toolkit (BOC Europe, 2015). After the modelling has been finished the process models are exported to XML files and uploaded onto the ProKEX Process Model Repository (PPMR). As a next stage these process models are loaded onto the ProKEX application as can be seen in Fig. 12.

After the process models in XML format are downloaded and the model that we wish to work with is selected, the application loads the job roles from the XML representation of the organizational diagram as shown in Fig. 13, Stage 4. In this example the insurance (CIG) process model was selected.

In the next stage the job role for which the analysis is carried out has to be selected as the Concept Group will be created in a job role basis. As a job role is selected, the tasks—to which the employee with the given job role is connected as

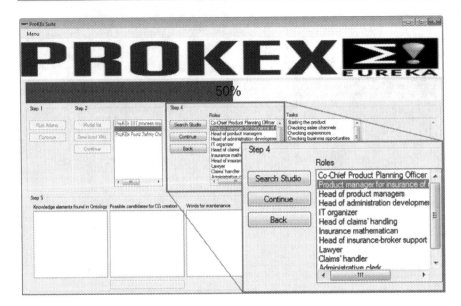

Fig. 13 Product manager for insurance of real estates is selected

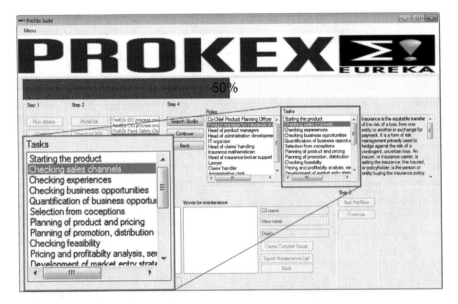

Fig. 14 Task list where Product manager for insurance of real estates is performer

an executer—are listed as well. The task descriptions can also be browsed in the application Fig. 14, Stage 4.

By clicking on the 'Search Studio' button the descriptions of the tasks where the employee with the selected job role is the executer are gathered into one corpus, and

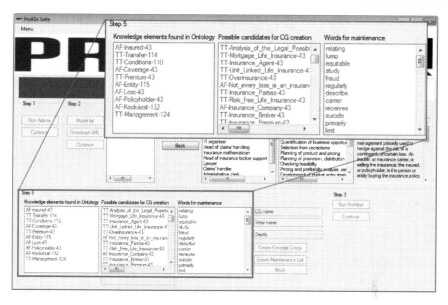

Fig. 15 Concepts from the Studio System in connection with the selected role

the text mining is carried out on it. In this way the list of phrases is produced that will form the basis of the ontology tailoring process. Based on this list a search in the Studio ontology is carried out. The ontology concepts are fully or partially identified by the phrases, though some elements of the list, which is the product of the text mining, may not match with any ontology node at all, as outlined previously. In this way the application creates three lists as shown in Fig. 15, Stage 5.

The listbox named the 'Knowledge elements found in the Ontology' contains the fully-identified ontology concepts, and the listbox named as 'Possible candidates for CG creation' the partially identified ontology concepts. The 'Words for maintenance' listbox contains those phrases which do not match any knowledge elements in the STUDIO. The nodes identified fully or partially in the ontology will serve as the basis for further tailoring process. The phrases which did not refer to ontology concepts can be exported as a list, and can be sent to the knowledge engineer, who is able to decide whether they represent a relevant knowledge element. In this way they have to be classified into the ontology, or not. The fully identified nodes will form part of the further work mandatorily, while the knowledge engineer is allowed to determine which nodes of the partially identified ontology concepts should be used, as shown in Fig. 16, Stage 5. The application suggests a name for the Concept Group and for the View which also can be modified by the knowledge engineer. The extent of the depth used during the tailoring process can also be set in this stage.

In this example fully identified nodes are, for example, the 'Insured', 'Transfer', 'Conditions' ... or the 'Management'. The prefix of the elements indicates the class which the given node belongs to in the STUDIO ontology. In this example 'TT'

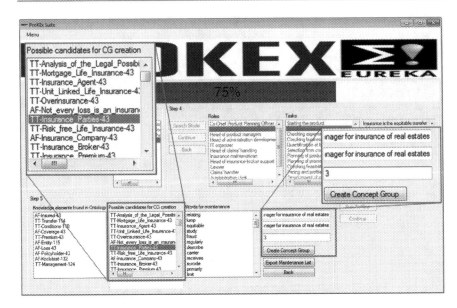

Fig. 16 Additional concept is assigned to the job role knowledge

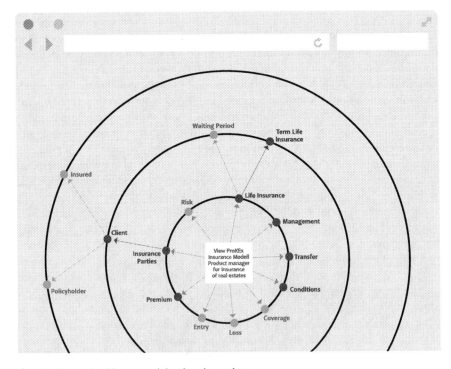

Fig. 17 Customized Insurance job role sub-ontology

means 'Knowledge Area', and 'AF' means 'Basic Concept' (Vas 2007). For more information about STUDIO classes see chapter "STUDIO: Ontology-Centric Knowledge-Based System". In Fig. 16, the node 'Insurance Parties' is selected by the knowledge engineer, the name of the Concept Group and the View is set to the name of the job role, namely 'Product manager for insurance of real estates' with a prefix which indicates the class of the object created. In Fig. 16 it can also be seen that the extent of the depth measure was set to three. By clicking on the 'Create Concept Group' button the tailoring algorithm is activated and the Concept Group is created in the STUDIO system. The Concept Group created on the basis of the data in Fig. 16 is represented in Fig. 17.

As shown in Fig. 17 the hierarchy given by the domain ontology is retained during the ontology tailoring process. The central element of the graph is the View object, which is the Start node of the Concept Group. At the first level (inner circle) those nodes are shown, for which no relevant neighbor could be found within the given distance (3) along the way to its Curriculum elements. In other words these nodes have no parent node related to the given topic, which is now outlined by the knowledge elements found in connection with the 'Insurance Process Modell Product manager for insurance of real estates' job role. After the first level the ontology hierarchy and structure was retained, for example the 'Insurance Parties' node is in a 'Requires Knowledge Of' relation with the 'Client' node.

As demonstrated in this chapter creating the meta-structure (Concept Group) is the goal of the tailoring stage of the ProKEX Suite. In the STUDIO system at least one question and piece of learning material may have connected to all the nodes of a Concept Group. In this way the Concept Group is a structure for application oriented knowledge tests, and in this example it is the skeleton of the test for the 'Insurance Process Modell Product manager for insurance of real estates' job role. In other words the testing algorithms work based on these meta-structures (Weber & Vas, 2015). Chapter "STUDIO: A Solution on Adaptive Testing" provides an additional insight into STUDIO and the Adaptive Testing.

References

BOC Europe. (2015). *Adonis Business Process Management toolkit*. http://www.boc-eu.com/
FAL Labs. (2010). *Tokyo Cabinet: a modern implementation of DBM*. http://fallabs.com/tokyocabinet/
Gábor, A., Kő, A., Szabó, I., Ternai, K., & Varga, K. (2013). Compliance check in semantic business process management. In *Lecture Notes In Computer Science 8186 LNCS*. pp. 353–362.
Gillani, S. A., & Kő, A. (2014). Process-based knowledge extraction in a public authority: A text mining approach. In: Electronic government and the information systems perspective. Springer International Publishing. LNCS 8650 (pp. 91–103).
Ternai, K., Török, M., & Varga, K. (2014). Combining knowledge management and business process management–A solution for information extraction from business process models focusing on BPM challenges. In: *Electronic Government and the Information Systems Perspective*. Springer International Publishing. LNCS 8650 (pp. 104–117).

Vas, R. (2007). Educational ontology and knowledge testing. *The Electronic Journal of Knowledge Management, 5*(1), 123–130.

Weber, C., & Vas, R. (2015). Studio: Ontology-based educational self-assessment. In: *Workshops Proceedings of EDM 2015 8th International Conference on Educational Data Mining, EDM 2015*, Madrid, Spain, June 26-29, 2015. Madrid, Spain (pp. 33–40).

STUDIO: A Solution on Adaptive Testing

Christian Weber

1   Continuous Training

Tell me and I'll forget; show me and I may remember; involve me and I'll understand.
Chinese Proverb

With the on-going integration of new, computerized devices in our daily lives and especially the development of new technologies accelerating with the push of the Internet, different, previously manual tasks have re-emerged as new technology enhanced versions. Training and education are especially profiting here from the new opportunities in integration and developments in the related fields of science.

Assessing human education, abilities and various aspects of performance always comes with the need for a strong set of methodologies from the field of education. They are supported by enhancements in the field of neuroscience, where more and more processes and relations of human reasoning are explored, enabling a better understanding of learning and integrating insights from the psychology of learning. At the same time, in the organisational context, the impact of a worker's knowledge is recognized as a central factor in the economic success of human capital, resulting in new disciplines on how to improve and assess individual knowledge. In the following the chapter will embrace the organisational view on continuous learning in the combination of learning and testing.

Kuckulenz (2007) introduces continuous training under the aspect of human capital. Human capital goes back to the seminal work of Becker (1964). Blundell et al. (1999) define the three main components of human capital as:

- early ability (acquired or innate)
- qualifications and knowledge acquired through formal education

C. Weber (✉)
Corvinno Technology Transfer Center, Budapest, Hungary
e-mail: cweber@corvinno.com

© Springer International Publishing Switzerland 2016
A. Gábor, A. Kő (eds.), *Corporate Knowledge Discovery and Organizational Learning*, Knowledge Management and Organizational Learning 2,
DOI 10.1007/978-3-319-28917-5_6

- and skills, competencies and expertise acquired through training on the job; (or acquired in the migration to a new job)

Learning, in the sense of human capital, is an investment in personal education, temporarily giving up a part of the current income in favour of an expected higher later income on a personal level and a higher expected productivity of the workers on the organizational level. Blundell, Dearden, Meghir and Sianesi (1999) addresses training, based on the use in empirical studies, as "generally defined in terms of courses designed to help individuals develop skills that might be of use in their job" and exclude formal training in the form of school and post-school education. Furthermore, they make evident a strong correlation between the performance in the two earlier components and the likelihood to engage in additional job related training and the performance on the same training.

Even though the general definition of training excludes regular education as formal training, embarking upon higher education implies a job relevant and job targeting decision and, while formalized, requires self-motivated learning behaviour. In this regard this chapter will address training and education synonymously, implying a process of for the job and on the job training and focus on these respects.

Furthermore, change is becoming a constant in daily life, following the pace and requirements of the markets, variations in labour and formal education undergoing constant change. Testing is the important factor to outline the current education and job- and work-horizon, supporting the knowledge worker to self-assess current strengths, potentials and the bottlenecks towards a next stage of education.

For a situation of training and education, computer aided, adaptive tests are one powerful enabler to connect the knowledge within a field of education. Current solutions to support a computer aided test preparation and test execution, neglect the fact that education and testing always takes place within a context, evolving around a problem context, given by the organisation, in line with the performance of the assessed individual.

This chapter will first provide an overview of the boundaries and transition points of learning and situations of learning, to motivate the vision of a context aware and context rich educational self assessment.

An introductory part about adaptive testing follows this and then addresses the ideas and potentials to adapt a testing and learning process to the learner, compromising between the need to educate and the ability to learn.

The final sections then shed light on the STUDIO approach on adaptive testing and will describe two complementing approaches for testing, introducing specific strategies to explore the knowledge of the learner on the conceptualization of the domain.

2   Formal, Non-formal and Informal Learning in the Organisational Context

Learning and especially organisational learning, as shown in chapter "Corporate Knowledge Discovery and Organizational Learning: The Role, Importance, and Application of Semantic Business Process Management—The ProKEX Case" is a fractured area with different views, strengthening a variety of differing fundaments and insights. As such the definitions of types of learning are still an object of discussion. With the target of creating a common understanding, different organisations over time promoted different sets of definitions, while matching in the three major types of learning: formal, non-formal and informal learning. Especially for use in policy and decision making, within and across countries, with the OECD, UNESCO and CEDEFOP, three grand multinational organisations proposed definitions for the three types of learning in line with the development in the literature of organisational learning, as shown below in Table 1.

All three definition streams share a common core with a slightly different emphasis, as seen in Table 2, based on the UNESCO definition and taken from (Werquin 2007). An important factor is the use and reasoning on the learner's intention of learning. Informal learning differentiate here that a situation may have an intention as the acting person has a goal but the main intention is not "learning" but rather learning is an inevitable, unconscious consequence.

Generalising and especially from an organisational view, formal learning connects to school and university education, offering certified degrees from accredited studies, while non-formal learning means organisational training activities which are "structured" and "organised" in their implementation but not formal in terms of the mediated education. This is independent of the potential that organisations (internal or external) can certify trained skills, with or without additional evaluation.

As the understanding of formal, non-formal and informal education is strongly connected to the cultural background and the spatial environment, the line between the former cannot be drawn in an absolute fashion and adds to the on-going discussion within the field. According with the variety of situations of learning, rather than being an absolute categorization it is a transient scale, as shown in Fig. 1, and is always a mix, depending on the specific situation.

Easterby-Smith et al. (2000) capture a part of the areas of on-going discussion under the term of operational learning. They name the starting fields "units or levels of analysis" addressing the question whether organisational learning is the sum of individual learning or needs to be approached differently. Here Garratt argues that a small group of people could have a major influence on strategic decisions within an organisation and as such a small group of senior managers could give a good approximation of the thinking of an organisation (Garratt 1987).

Fiol and Lyles suggest the organisational structure and procedures directly affects the individual learning (Fiol and Lyles 1985), while Shrivastava collects that the knowledge of individual decision makers creates policies and procedures for organisations which are then embedded into the organisational structure and the organisational socio-cultural norms (Shrivastava 1983). Shrivastava summarizes

Table 1 Types of learning for policy making

	OECD (OECD)	UNESCO (UNESCO), based on (Commission 2001)	CEDEFOP (CEDEFOP 2011)
Formal Learning	Formal learning is always organised and structured, and has learning objectives. From the learner's standpoint, it is always intentional: i.e. the learner's explicit objective is to gain knowledge, skills and/or competences.	Formal learning occurs as a result of experiences in an education or training institution, with structured learning objectives, learning time and support which leads to certification. Formal learning is intentional from the learner's perspective.	Learning that occurs in an organised and structured environment (in an education or training institution or on the job) and is explicitly designated as learning (in terms of objectives, time or resources). Formal learning is intentional from the learner's point of view. It typically leads to validation and certification.
Non-Formal Learning	Mid-way between the first two, non-formal learning is the concept on which there is the least consensus, which is not to say that there is consensus on the other two, simply that the wide variety of approaches in this case makes consensus even more difficult. Nevertheless, for the majority of authors, it seems clear that non-formal learning is rather organised and can have learning objectives.	Non-formal learning is not provided by an education or training institution and typically does not lead to certification. It is, however, structured (in terms of learning objectives, learning time or learning support). Non-formal learning is intentional from the learner's perspective (Werquin 2007).	Learning which is embedded in planned activities not explicitly designated as learning (in terms of learning objectives, learning time or learning support). Non-formal learning is intentional from the learner's point of view.
Informal Learning	Informal learning is never organised, has no set objective in terms of learning outcomes and is never intentional from the learner's standpoint. Often it is referred to as learning by experience or just as experience.	Informal learning results from daily life activities related to work, family or leisure. It is not structured (in terms of learning objectives, learning time or learning support) and typically does not lead to certification. Informal learning may be intentional but in most cases it is non-intentional (or 'incidental'/random).	Learning resulting from daily activities related to work, family or leisure. It is not organised or structured in terms of objectives, time or learning support. Informal learning is in most cases unintentional from the learner's perspective.

Table 2 Differences of learning frames (Werquin 2007)

	Organised	Learning objective	Intentional	Duration	Leads to a qualification
Formal Learning	Yes	Yes	Yes	Rather long and/or full-time	Yes[a]
Non-formal Learning	Yes or No	Yes or No	Yes or No	Rather short, or part-time	No[b]
Informal Learning	No	No	No	NA	No

[a]"Almost always"
[b]"Usually no"

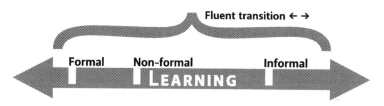

Fig. 1 Fluent transition between types of learning

further four main aspects of organisational learning: organisational learning as "adaptation", "assumption sharing", "developing knowledge of action outcome relationships" and "institutionalized experience", highlighting in his summary that organisational learning is partially motivated to respond to change.

People learn in a variety of different environments and situations and approach learning intentionally and unintentionally, following conscious and unconscious goals alike. Non-formal training in organisations usually takes place in a fit between the organisational goals and the interests of the workers, e.g. within the framework of seminars and training systems. Non-formal learning approaches are opening the potential for an effective and on-going education within organisations and make use of previous formal education while the teaching offered is driven by the organisational interest, also embedding as such the organisational culture and benefiting from the job relevance for the workers, and overall facilitating further implicit informal-learning processes.

Vaughan and Cameron (Vaughan et al. 2009) gives a set of reasons why assessment in the workplace is yet underexplored in the current literature, placing emphasis on the impact of informal learning, including one emphasis on the understanding of learning

> There are several, interwoven reasons why there is very little literature dealing directly with workplace assessment. Firstly, workplace assessment is closely related to forms of learning that are not recognised or understood as learning, which means that such learning is less likely to be assessed.

A technology enhanced testing and learning solution could act here as a great benefactor and yield a high potential for more additional informal learning outcomes. It could provide the impact of organisational learning in a timely and spatially independent fashion, based on the need of the organisation and accessible in situations where workers are searching for further education. Here the ProKEX approach fosters the hand-shake between organisational and workers learning interests through basing the learning and testing packages on the processes of the organisations which connect the structural aspects of the organisation with the job roles and job role competencies of the individual workers. To assess these competencies across a wide range of individuals a degree of adaptivity is mandatory for a working and sustainable assessment.

3 Adaptive Testing

Unless you try to do something beyond what you have already mastered, you will never grow.
Ronald E. Osborn

In knowledge driven professions adapting to new fields of knowledge and expertise is a daily necessity, which also covers cases of organisation-internal change. Maintaining and extending here the right set of knowledge, requires personalized development with the potential to follow multiple paths of education in dependency of the personal context. To fully unlock the personal potential in a resourceful way, education has to mediate between the need to educate and the ability to learn. This requires an open and adaptive approach to testing, maintaining a balance between the time to assess the state of the learner and his/her time to learn detected gaps in education.

Among the different approaches for adaptive testing, computerized adaptive testing is a well explored example on how to realize adaptivity and provides a solid researched ground with a strong link to the initial idea of adaptive testing.

3.1 Computerized Adaptive Testing

The first adaptive tests for assessment go back to Binet, who created a test, asking questions in regard to human ability (Simon and Binet 1904), adapting the difficulty of the test to the performance of the test candidate. A methodology which was refined by Lord (Lord 2012), Henning (Henning 1987), Lewis and Sheehan (Sheehan and Lewis 1992) and Reckase (Reckase 1974), leading to a first computerized adaptive test (CAT), was later extended by the models of Rasch (Rasch 1960) and Wright (Wright 1988).

An adaptive methodology was first applied in psychology. Alfred Binet's goal was to develop intelligence tests which were aimed at diagnosing the individual. By moving away from the target of assessing an assumed homogenous group he could eliminate the issue of fairness, providing the same test framework but assessing it on the basis of personal performance. He then customized the test based on an individual rank, by ordering the items (questions) according to their assumed difficulty. He would then start testing the candidate through continuously estimating the probable level of the candidate's ability, based on the difficulty of the asked items.

The selection of items then represented a subset of the overall set of available question items which he believed to meet the detected level of the candidate. If questions were answered correctly Binet would assess successively harder item subsets and would stop as the candidate failed frequently. Vice versa, if the candidate failed questions, then Binet administered successively easier item subsets, finishing if the candidate succeeded frequently.

For an adaptive test, in contrast to a classical linear test, the number of test questions and their order is determined within the process of testing. The target is to find a trade-off between determining the knowledge level of the test candidate as precisely as possible while administering questions in numbers as low as possible. An adaptive tests favours detecting the capabilities of the test candidates in contrast to linear tests, which are constructed to reflect and meet the requirements of test-givers.

A challenge for this type of adaptive tests is, concurrent to linear tests, to set up a test or set of test items which assess an appropriate level of difficulty. Both an easy test and one that is too complex may fail. If a test is too easy for a candidate it may invite unwanted human behaviour such as careless mistakes or unneeded, yet wrong assumptions. Furthermore, questions and question sets which are much too hard to answer produce generally unreliable test results, influencing candidates to give up early and inviting frustration, leading to guessing, response patterns and additional unwanted and biasing behaviour. Within traditional testing methods, candidates should be assessed by the same set of questions, static in number and assessed completely. This renders the results and the performance of each candidate comparable and reliable for repetition

Adaptive testing takes a different approach here: As each candidate may receive entirely different questions with a changing order of questions the result of the test is determined by mapping the detected difficulty level of the candidate to an assumption of the candidate's ability level (the assessment will always include a certain amount of error). Pictured as an algorithm the answers to (previous) questions determine the next questions. The stages in Fig. 2 picture the progress of the algorithm. The CAT, in this respect proposes main characteristics, recorded in more detail in chapter "STUDIO: Ontology-Centric Knowledge-Based System", which are also appropriate for a range of different adaptive tests (Table 3):

CAT is considered as a proven solution, providing tests which are adaptive to the ability of the learner, while optimizing time and precision of the testing (Triantafillou et al. 2008; Čekerevac Zoran and Petar 2013). However, fundamental implementations of CAT do not take into account that education and testing is taking place in a context—a gap which is tackled by the STUDIO implementation.

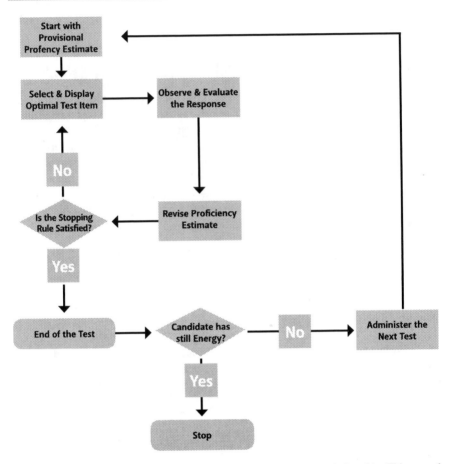

Fig. 2 Computerized Adaptive Test algorithm, following the process designed by Thissen and Mislevy (1990)

Table 3 Characteristics of adaptive tests, revisited

Characteristic	Description
Time independent	Tests can be taken any time.
Minimizing copying	Tests are tailored to the learner and are not identical.
Improved concentration	Questions are presented one at a time.
Strict assessment	Questions are: fixed after answering and cannot be skipped.
Dynamic ending	The end of the test depends on the answers of the examinee.

3.2 Human Centred Testing and Training

Following the on-going discussions of the potential impacts and lessons learned from the European Pisa studies on student performance, among other mirrors of education, the Pisa study is one proxy for the observation that the European educational system is in a situation fraught with tension, searching for ways to change. Education in many nations is still conservative, using frontal and static teaching methodologies while the content is focused on a local or national canon, neglecting a more personal education, aligning to the different potentials of the students. The early stages of formal education as such tend to create a middle population of future employees, which share a similar set of background knowledge. Besides providing the benefit of a standardized entry into education and the labour market, the current formal system, staying static, doesn't focus sufficiently on developing the inherent potentials and abilities of students and later workers.

Contrasting, throughout all markets, organizations have to change faster and more rapidly. An increasing amount of products become high technology creations, rendering the design, development and production more complex. Resulting, organizational changes are becoming proxies for new market and technology requirements, passing down the requirements, with a time shift, to the different job roles workers have to fulfil in order to cope with their revising daily tasks. While new technologies and production processes are continuously developing and extending to enable a better adaptability to changes in product requirements, workers are increasingly in conflict with the static learning culture, with their previous inflexible education and the increasing need for adaptability on the job.

Technology here has the potential to bridge the gap between a static educational culture and fast changing job roles through making use of methodologies to adapt to the individual learners - a static, generalized education meets advancing and dynamic technologies. The long term goal is to offer a new flexible training service which is easy to adapt to the needs of the organization and the capabilities of the workers. Self-learning and learning under human support offers the ability to modify the framework, order and medium of communication and learning.

Yet, transforming the daily-life process of learning partially into a computerized framework sacrifices a part of the freedom of learning. Within the ProKEX approach this is not seen as a limitation but as a part of the strategy to create a framework for the learning and testing experience.

The testing of job knowledge in relation to job roles which are needed for specific organisational processes comes with an interest in applying strict test- and learn-frameworks which follow the processes rather than enabling an open exploration of the involved concepts.

In this regard the creation of a solution is process-centred. In contrast, presenting the results of assessment in the context of the related concepts, while enabling the employee to extend their understanding of the process-related knowledge in cycles of testing and learning is human-centred. It is human-centred as it depends on the performance of the worker in the STUDIO system and profits from the adaptability of the system to the individuality of each single user.

A continuous testing and the visualization of the mastered process knowledge leads over time to three areas of impact, as shown in Fig. 3, which the system unlocks by making use of an enhanced adaptive assessment solution:

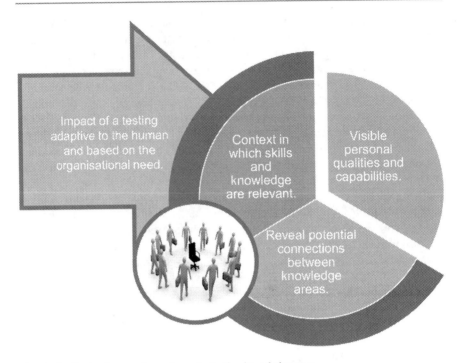

Fig. 3 Qualitative impact dimensions of adaptive knowledge tests

1. Grasping visually the progress of the personal testing makes it possible to achieve an understanding of personal qualities.
2. Secondly, the provided relations of relevant knowledge areas are a reflection of process-required knowledge and its structure. The consecutive mastering of areas on the other hand illustrates the understanding of the individual. The visualization here yields the potential to discover new relationships between knowledge areas if specific areas are mastered regularly in conjunction with other knowledge areas.
3. Thirdly, assessing concepts needed to perform specific processes sets the concepts into the context of required skills and knowledge and enables a broader understanding of the job roles in which they are relevant.

4   The STUDIO Approaches on Adaptive Testing

The assessment in the ProKEX framework takes into account two visions on assessment, resulting in two alternative approaches for testing:

The priorities of the first, top-down driven algorithm, is the fast, strict and sharp detection of the "job role knowledge" to find learning potentials to improve the job role related skills. The algorithm strictly follows the structure of the domain, top-down which resembles the logic of the processes and their required knowledge areas, detailing down to job roles and task related knowledge.

The second, bottom-up driven algorithm prioritizes to trial on the variety of task related knowledge on one hand and the detailed knowledge to perform the available bandwidth of tasks on the other hand, starting from the detailed bottom of the domain structure. This enables a better overview of the individual abilities of the employees to then explore the potentials of the learners. It combines the learned lessons by following the structure to the "top" through more aggregated knowledge elements, while implicitly evaluating which job roles could be performed, based on the tested detailed task related knowledge.

4.1   The Prerequisites of Adaptive Assessment and Assessment Path Creation

The creation of each new self-assessment test begins with the interaction of the tutor with the system. The rational of this initial stage is to base the test on the expertise of a real world expert in terms of selection and based on a tutor driven estimation of an appropriate granularity between the knowledge area levels for the learner. Within the ProKEX framework this expert driven selection is completed by a process driven extraction of knowledge areas, relevant for the organisational processes.

To create a regular self-assessment test through STUDIO, the tutor needs to select the relevant knowledge areas and connect them to concept groups which together create a tree of groups. Within the ProKEX framework this takes place based on the support of the process model. The resulting tree then forms a picture of a sub-ontology of the main domain ontology. The specifics of this first part of the process are detailed in more depth in (Neusch and Gábor 2014) and the corresponding chapter "Ontology Tailoring for Job Role Knowledge". For each concept group the system will import related knowledge elements from the domain ontology and complete the test framework. This extraction stage completes the framework with all the knowledge areas and relations from the domain ontology which are necessary to connect the already selected knowledge areas, based on the concept groups.

The output of the extraction is a cached directed graph representation of the modelled assessment domain. By definition, the top element of the topmost concept group will be set as the start-element and root of the tree shaped graph. The start-element acts as a fixed point from which the top-down assessment algorithm will start and to which the bottom-up algorithm will explore the graph. As such the start-element is the centre from where the imported knowledge structure is interpreted for testing.

In the next stage the selected assessment algorithm will start and move through the knowledge structure, based on the internal navigation rules while online administering the questions connected to each knowledge area the algorithm

Table 4 Necessary assumption for traversing the knowledge structure for assessment

Assumption	Description
Ordering	All knowledge areas are connected with "part_of" and "requires_knowledge_of" relations. The result is that every path, starting with a start-element, will develop on average from general concepts to detailed concepts. By implication, any methodology to select concept groups for the test definition has to be designed in a way that it selects and orders following concept groups accordingly to also lead from general at the top to more detailed groups at the bottom of the structure.

selects. To explore the knowledge structure for both internal test algorithms, the system makes use of one central assumption, depicted in Table 4:

Each created assessment-test structure defines a sub-ontology of the source domain ontology. This extracted blueprint of the test starts with the start-element of the highest defined concept group. To load and complete the knowledge structure for the self-assessment, the system follows the following stages in a cycle in order to load the structure, based on the test definition described in the previous section:

1. Knowledge-elements are connected through relations. Each relation type between two knowledge-elements has one unique direction, fixing the extracted tree as a directed graph. The system will load all relations between two knowledge-elements, which start with the start-element and ends on another knowledge-element. This creates a two level structure where the start-node is a parent-element and all related, loaded elements are stored as child-elements.
2. The algorithm then successively selects each child-element of the start-element and defines it as a start-element in its own process.
3. When no more knowledge-elements for a parent-element can be loaded, the sub-process stops.
4. The system then repeats the first stages till all knowledge-elements are loaded into the created tree-structure. When all the sub-processes have stopped the knowledge structure has finished loading.

The overall process flow is shown in Fig. 4.

4.2   A Top-down Approach on Testing

To ensure a working top-down assessment, the system must satisfy the assumption that missing knowledge at an early stage of the knowledge structure hierarchy disables the learner from answering questions for knowledge areas on more detailed levels. This assumption is summarized in Table 5 and follows the taken process view on the assessment that the knowledge about the high level processes, which correspond to higher levels in the knowledge structure based on the process extraction, is needed to perform to an acceptable level in job roles and tasks, modelled deeper in the knowledge structure.

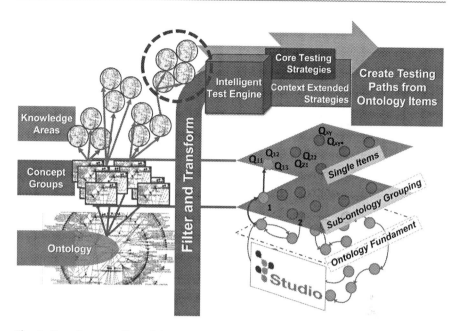

Fig. 4 Overall process flow of the concept extraction from domain ontology, based on concept groups

Table 5 Necessary assumption for traversing the knowledge structure in a top-down setup for assessment

Assumption	Description
Top-down knowledge dependency	If a test-candidate fails on more general concepts the system will assume that he/she will also fail on more detailed concepts. Furthermore, when a sufficient number of detailing concepts failed, the parent knowledge will not be necessarily covered and will be classed as having failed, too

Based on the first assumption, defined in the previous section, the deeper a knowledge-element is within the tree, the more detailed the concept of the element is, creating a hierarchy going from general concepts to specialized concepts while moving down the tree structure. Out of the directed one-directional nature of the defined relations, this loading-process provides a directed tree of the knowledge structure to the test algorithm. An example visualization of a tree is shown in Fig. 5.

Adapting to the tree shaped knowledge structure, the top-down assessment triggers the following stages to assess the represented knowledge, based on the questions connected to each knowledge-element:

1. Beginning with the start-element, the test algorithm activates the child knowledge-areas of the start element.

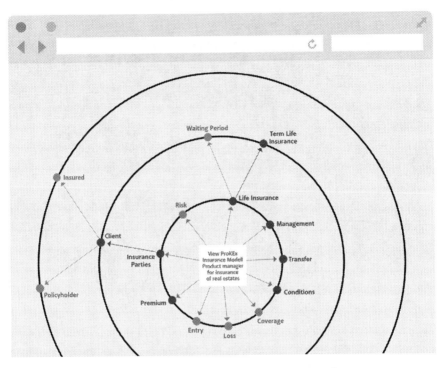

Fig. 5 Excerpt from the sub-ontology visualization, showing the desired tree structure

2. The top-down algorithm then selects the first child-knowledge area and extracts a random question out of the question storage connected to the knowledge-element and assesses the answer from the learner through the testing interface.
3. When the learner fails the assessed question, the algorithm marks the element as failed, or else it is marked as passed.
4. The system then selects the next non-failed knowledge area, accessible directly or through passed nodes from the start-element, promotes it as a parent node and queries a random question from it to repeat the process afterwards.

Following the test algorithm in a cycle, the system dives the knowledge structure top-down and continuously triggers more questions depending on the learner's answers and on the designed and extracted model of the relevant education. In this regard the STUDIO system adapts the test to the performance of the learner, based on the answers to the test questions. This triggers the choice of following or not following more knowledge elements on the same branch of the knowledge tree. As this assessment works on the structure, created on the basis of the extracted organisational processes, the system adapts the testing of the domain knowledge to the assessed knowledge of the learner, providing as such an adaptive solution for the self-assessment.

This process of mapping the learners performance to the conceptualisation of the domain ontology, resembles the concept of overlay based student modelling (Chrysafiadi and Virvou 2013). While the learner continues to use the self assessment through phases of testing and learning, to evaluate personal knowledge, he/she will dive down further into the knowledge structure and continuously explore more and more detailed areas of the target education.

As the approach is sufficient to assess the alignment of learners to the defined organisational processes this approach is limited in regard to the speed of exploring single knowledge elements. The algorithm for assessment stops following a branch of the knowledge structure as soon as the learner fails on a knowledge-element. Especially failing on top level elements near to the original start-element, results in the early exclusion of a complete sub-areas of the overall knowledge domain in the testing process. This can lead to iterations where after a short series of failed answers the assessment will stop on the first level of knowledge-elements, giving no specific feedback on learning and missing the opportunity to assess single task related knowledge, which may draw a picture of the current capabilities of a learner.

This correlates with the desired behaviour fields of testing as computerized adaptive testing (CAT) (Linacre 2000; Edwin Welch and Frick 1993), but it ignores the possibility that learners could yield knowledge of more detailed concepts, while not having fully mastered general concepts or yet miss the assets to understand the questions of higher level concepts.

4.3 Bottom-up: A Path Based Assessment and Self-Assessment

In a bottom-up assessment, in contrast to the top-down testing described earlier, the assumption is that learners will know details about the represented domain, even if they cannot answer questions for high level concepts. A failing on earlier concepts may be hidden in a yet missing comprehension of correlations and consequences, which is reflected in the differentiating dependency assumption, phrased in Table 6.

Further questions for high level knowledge elements may be considerably harder to phrase and create as they have to represent a trade-off between size, concept dependencies and the numbers of concepts needed to make a statement about the core of a concept and its implications. As such the probability for flawed or biased questions on higher levels is higher than for detailed concepts.

Taking a process-related view, learners may have sufficient knowledge to fulfil tasks of the target job roles, attached to the processes, but may lack a higher level understanding of the reasoning behind processes. Furthermore, they may already have an understanding of target processes, paired with the power to apply it in everyday situations, appropriate to the classification of the blooms taxonomy (Krathwohl 2002), but yet lack the analytical proficiency to transform them into a specific question.

As such there is a rational, especially in early learning cycles, to start to assess more detailed knowledge first to create an understanding of the current skill level and the compliance to processes.

Table 6 Necessary assumption for traversing the knowledge structure in a bottom-up setup for assessment

Assumption	Description
Bottom-up knowledge dependency	If a test-candidate fails on more detailed concepts the system will assume that he/she will also fail on more general concepts. If a test-candidate fails on more general concepts he/she could potentially still succeed on more detailed concepts. The result is that each knowledge-area or element is relevant for the main test goal. As higher level concepts are comprised of an aggregation of detailed concepts, sub-level concepts have to be tested as completely as possible to explain failure at higher levels.

Furthermore, if the assessment stops in the early stages this may discourage learners to retry the test after gaining more insights. This then prevents further exploration of the knowledge structure and acquiring insights into the missing knowledge. A learner may not have the overall knowledge of an educational area but it can prove to be vital to know proficiency on preconditioned knowledge areas.

A solution to explore the knowledge of learners broader and more detailed in a bottom-up approach, while still keeping an evaluation framework comes in the form of the creation of assessment paths. Assessment paths are a generalization of the concept of connected knowledge elements and describe paths through the knowledge structure which connect one knowledge element to the respective start-element. A path can thereby include an unlimited amount of intermediate knowledge elements which are needed to connect to the start-element. To prevent loops in the directed graph, the final algorithm makes use of black-lists of visited nodes, combined with a backtracking algorithm to continue to create and explore alternative paths.

To enable the new path concept, the assumptions about the structure, based on the top-down algorithm, have to be modified and extended, resulting in the assumptions phrased in Table 7. In the top-down approach, as the algorithm starts from the start-element, each passed element in an assessment would be connected with a path of passed elements to the start-element.

As the bottom-up algorithm starts from bottom knowledge-elements, a path from a passed element to the start-node may include failed elements. To cope, testing and evaluation are divided for the bottom-up assessment, as shown in Table 7. Following this, passed elements will only be accepted if they are connected to a path of other passed elements, connecting without interruption to the start-element.

With the new concept of assessment paths, the tree shaped knowledge structure extracted for the assessment, describes a set of possible paths from each knowledge-element to the start-element. These paths represent routes through the knowledge structure and traces how far a learner masters connected concepts. They show further how complete the capabilities to fulfil single job roles and process requirements are, while the set of succeeded paths together mark knowledge-areas in which the worker excels. Beside the integrated feedback on the depth reached within the structure, which is also available through the top-down

Table 7 Revised assumptions for a path based assessment

Assumption	Description
Extended Ordering	All knowledge areas are connected with "part_of" and "requires_knowledge_of" relations. The result being that every path, starting with a start-element, will develop on average from general concepts to detailed concepts. To sufficiently explore the knowledge structure, for each set of knowledge-elements, accessible through a path of connected relations, the test will first select knowledge-elements which have the highest amount of intermediate knowledge-elements.
Path/Knowledge evaluation assumption	If a test-candidate fails in more general concepts he/she could potentially still succeed in more detailed concepts. The result being that each knowledge-area or element is relevant for the main test goal, if there is a path of knowledge-elements to the start-element, which includes only knowledge-elements marked as passed.

approach, the bottom-up assessment furthermore yields the potential to group assessed knowledge based on the set of finished and unfinished paths.

The trade-off of the use of paths comes in the shape of an additional algorithmic impact. The loading and creation of paths has to be partially pre-fetched and based on the complexity of the structure a multitude of different paths could be complex and more expensive in extraction than the top-down approach. To lessen the strain on resources, the new algorithm makes use of the knowledge on the evaluation of paths. Only paths between a node and the start-element which include no failed knowledge-element will be evaluated. The result is that each failed knowledge-element leads to an automatic "block" and cut-off of the current path above the failed element. As such every alternative path of a cut-off element, which has no other path to the start-element will be omitted. So while the algorithm explores the knowledge structure it successively decreases the set of unexplored paths that will not be assessed for the evaluation.

Figure 6 shows the creation process of paths of the bottom-up algorithm. The input comes out of the concept group-based extraction process, as described in the previous sections. The resulting sub-ontology delivers the structure which is used to run the assessment.

The path creation over the course of the overall assessment will continuously trigger these three stages:

1. The stage "Build" extracts fitting knowledge-elements from the knowledge structure, starting from the start-element and combines them into paths.
2. "Use", triggers the central assessment which then assesses the path based on the connected questions from the bottom to the top element.
3. The stage of "Store results" is a concurrent process, storing the success or failure of knowledge elements and cuts off elements which are marked as failed, based on failed questions.

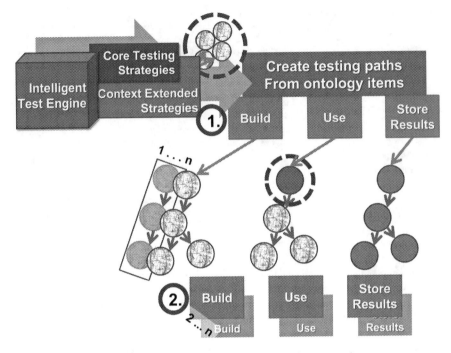

Fig. 6 Path based assessment process, succeeding the sub-ontology extraction process

So if any elements fail, they will be marked as failed and essentially block the later part of the current path to the start-element. With this cut-off procedure the algorithm minimizes the set of future sub-paths to assess. As designed, the system accepts every path of knowledge-elements, which reaches the start-element through offered relations and passed knowledge elements. As a result a failed knowledge-element divides a path into a "top" part which could still reach the start-element and a failed bottom part that won't be considered for the final result.

Figure 7 shows the result summary visualization of the bottom-up self-assessment test. Red/dark dots signal knowledge-elements which the learner failed, while green/light dots identify knowledge-elements which have been passed. The image visualizes the potential to reason on cleared and not yet cleared areas of the domain. While some concepts are known, higher level knowledge-elements could not be passed and mark sections for further learning.

With the assessment of detailed knowledge-elements, failed, higher level knowledge-elements could be broken down into areas of missing knowledge. This additional information would, within the top-down approach, only become known for quite completely explored tests and may increase the number of learning iterations for mastering the domain knowledge. In contrast to these additional learning profits, a single bottom-up test takes considerable longer in earlier testing iterations than a top-down test.

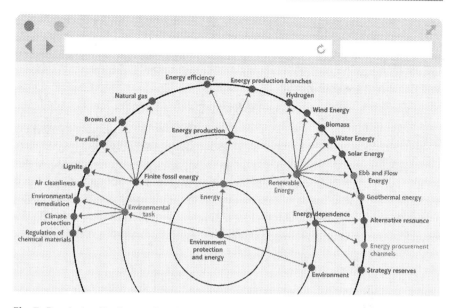

Fig. 7 Result visualization as educational feedback for the learner, based on the bottom-up testing alternative

The choice of a test algorithm for a specific assessment goal has to include an initial analysis of the requirements of the assessment. In cases of a large scale selection of well prepared learners for the assignment on specific job profiles the strict top-down testing scheme is more suitable and as it more strictly covers the connected organisational process. For pre-filtered groups of candidates the bottom-up assessment may provide a wider profile of the capabilities of each individual and enable a more informed selection decision.

5 Future Work

Within an adaptive approach to testing, the test adapts to the short-term performance of the learner. The collection of mastered and not yet mastered knowledge areas draws a map of the current learning state of the test candidate. Tracking the candidate's performance and continuing the test in knowledge areas which are related in terms of the structure of the domain and the feedback of the learner alike provides an adaptive test experience. Yet the underlying domain and the test framework is unaffected by the information gain generated from the test feedback. The next generation of adaptive test algorithms will make use of this additional source of information. To do so it needs a new middle layer between the domain ontology and the test engine, providing the test experience for the learner. This layer will have a two-fold manifestation.

From the scientific point of view an additional layer of labels is beneficial, expressing additional information about questions and learning material, such as their suitability for specific behaviour patterns or the frequency of failing in testing situations. A suitable scientific concept in this case is the overlay model from the field of student modelling (Chrysafiadi and Virvou 2013). This goes together with the need for a methodology to fill, complete and manage labels and feed them into the testing process at runtime. A possible approach for making use of labels in testing for known behaviour patterns is addressed in detail in (Weber et al. 2015).

Managing additional information labels requires monitoring the development and changing of desired profiles and is strongly connected to a monitoring of the assessment with a focus on the generated test feedback. Picturing the analytical stages of detection and matching a software conceptualization of the used methodologies will be needed. A resulting analytical framework based on the generated insights will be the foundation to shift the manual process of investigation and profiling into an automated system, completing the gathered flow of information in the background and giving access to the created source to the assessment algorithm.

Today the test changes the test candidate, enabling a personalized learning experience and providing the right set of learning materials to progress to a better state of knowledge and understanding. In the future the test candidate will also change the test in a complete change and feedback cycle, as seen in Fig. 8, and improve the adaptation potential. More knowledge about the test candidate enables

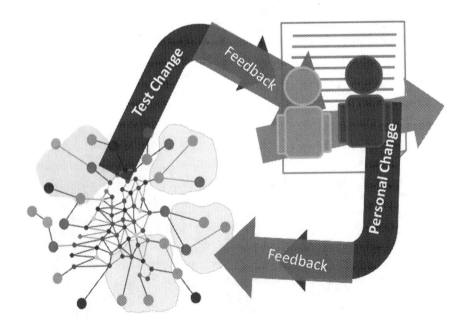

Fig. 8 The next stage of adaptivity: The test changes, based on the long term performance feedback of the learner

a better matching of test questions and learning material and this enhances the potential to learn faster and in a more sustainable way as the new knowledge becomes better embedded into existing lessons learned.

Potentials are already visible. Detecting learning styles and learning preferences gives the power to profile the test candidate on different dimensions of behaviour. While scientific proof of an acceleration of learning, based on pure learning style detection is still lacking (Pashler et al. 2008), a better fit of material and learner has the potential to increase the motivation to learn and raise the consciousness of the learning process. Detecting general level profiles such as a cultural context, could further enable the reduction of known biases in test questions (Makransky and Glas 2013) and improve testing in terms of adapting closer to the inert ability of each learner.

6   Conclusion

As Vaughan and Cameron (Vaughan et al. 2009) address in the context of informal learning: "... workplace assessment is closely related to forms of learning that are not recognised or understood as learning, which means that such learning is less likely to be assessed." The ProKEX approach on process based assessment can make it possible to overcome the assessment gap for domains with a strong informal learning culture by assessing workers on the shared conceptualization of processes.

Adapting the assessment strategy to the capabilities of the learner and the requirements of the field of assessment helps to create a clearer picture of the current state of learning and provides each learner with visual and process-related feedback on how to comply with the process requirements. Furthermore a better picture of current knowledge gaps provides clear recommendations on the next stages of learning. While this raises the awareness of learning and makes it possible to shift more parts of the informal learning into a conscious and structured process of learning, it provides a working and sustainable approach on a grand scale to non-formal learning.

Together with the future potential of an active profiling system, further extending and improving the aspect of adaptivity and individual progressing, the implemented assessment approach gives organisations the possibility to select and pre-select workers process-centred and process-related. When entering the job, the workers then have the opportunity to learn with the "hands-on" required knowledge in a continuous process of non-formal and organisational focused education.

References

Becker, G. S. (1964). Human capital: A theoretical and empirical analysis, with Special Reference to Education, by Gary S. Becker, ... London.

Blundell, R., Dearden, L., Meghir, C., & Sianesi, B. (1999). Human capital investment: The returns from education and training to the individual, the firm and the economy. *Fiscal studies, 20*, 1–23.

CEDEFOP (2011). Glossary—Quality in education and training.

Čekerevac Zoran, A. S., & Petar, Č. (2013). Knowledge assessment and application of computer adaptive testing. *Knowledge Assessment and Application of CAT // MEST, 1*, 16–30.

Chrysafiadi, K., & Virvou, M. (2013). Student modeling approaches: A literature review for the last decade. *Expert Systems with Applications, 40*, 4715–4729.

Commission E. (2001). *Making a European Area of Lifelong Learning a Reality: Communication from the Commission.* Office for Official Publications of the European Communities

Easterby-Smith, M., Crossan, M., & Nicolini, D. (2000). Organizational learning: Debates past, present and future. *Journal of Management Studies, 37*, 783–796.

Edwin Welch, R., & Frick, T. (1993). Computerized adaptive testing in instructional settings. *Educational Technology Research and Development, 41*, 47–62. doi:10.1007/BF02297357.

Fiol, C. M., & Lyles, M. A. (1985). Organizational learning. *Academy of Management Review, 10*, 803–813.

Garratt, B. (1987). *The learning organization: And the need for directors who think.* London: Fontana Paperbacks.

Henning, G. (1987). *A guide to language testing: Development, evaluation, research.* Rowley, MA: Newberry House Publishers.

Krathwohl, D. R. (2002). A revision of Bloom's taxonomy: An overview. *Theory Into Practice, 41*, 212–218. doi:10.1207/s15430421tip4104_2.

Kuckulenz, A. (2007). Continuing training. In: Studies on continuing vocational training in Germany (pp. 9–52). Heidelberg: Physica-Verlag.

Linacre, J. M. (2000). Computer-adaptive testing: A methodology whose time has come. In J. M. Chae, S. Kang, U. Jeon, & E. Linacre (Eds.), *Development of computerized middle school achievement tests.* Komesa: Seoul.

Lord, F. M. (2012). *Applications of item response theory to practical testing problems.* Hillsdale, NJ: Taylor & Francis.

Makransky, G., & Glas, C. A. W. (2013). Modeling differential item functioning with group-specific item parameters: A computerized adaptive testing application. *Measurement, 46*, 3228–3237.

Neusch, G., Gábor, A. (2014). ProKEx—Integrated platform for process-based knowledge extraction. ICERI2014 Proceedings 3972–3977.

OECD Recognition of Non-formal and Informal Learning—Home—OECD. http://www.oecd.org/education/skills-beyond-school/recognitionofnon-formalandinformallearning-home.htm. Accessed 14 Jun 2015

Pashler, H., McDaniel, M., Rohrer, D., & Bjork, R. (2008). Learning styles concepts and evidence. *Psychological Science in the Public Interest, 9*, 105–119.

Rasch, G. (1960). *Probabilistic models for some intelligence and attainment tests.* San Diego, CA: MESA Press.

Reckase, M. (1974). An interactive computer program for tailored testing based on the one-parameter logistic model. *Behavior Research Methods & Instrumentation, 6*, 208–212. doi:10.3758/BF03200330.

Sheehan, K., & Lewis, C. (1992). Computerized mastery testing with nonequivalent testlets. *Applied Psychological Measurement, 16*, 65–76. doi:10.1177/014662169201600108.

Shrivastava, P. (1983). A typology of organizational learning systems. *Journal of Management Studies, 20*, 7–28.

Simon, T., & Binet, A. (1904). Méthodes nouvelles pour le diagnostic du niveau intellectuel des anormaux. *L'année psychologique, 11*, 191–244. doi:10.3406/psy.1904.3675.

Thissen, R., & Mislevy, R. J. (1990). Testing algorithms. In H. Wainer, N. J. Dorans, R. Flaugher, B. F. Green, & R. J. Mislevy (Eds.), *Computerized adaptive testing: A primer* (pp. 101–134). Mahwah, NJ: Lawrence Erlbaum Associates.

Triantafillou, E., Georgiadou, E., & Economides, A. A. (2008). The design and evaluation of a computerized adaptive test on mobile devices. *Computers & Education, 50*, 1319–1330. doi:10.1016/j.compedu.2006.12.005.

UNESCO The Learning Continuum. http://www.unesco.org/new/en/education/themes/strengthening-education-systems/quality-framework/technical-notes/learning-continuum/. Accessed 15 Jun 2015

Vaughan, K., Cameron, M., Aotearoa, A., et al. (2009). *Assessment of learning in the workplace: A background paper.*

Weber, C., Truong, H. M., Vas, R. (2015). Context-aware self-assessment in higher education. In: EDULEARN15 Proceedings (pp. 5910–5920). Barcelona, Spain: IATED.

Werquin, P. (2007). Terms, concepts and models for analyzing the value of recognition programmes. In: Report to RNFIL: Third Meeting of National Representatives and International Organisations, Vienna, October (pp. 2–3).

Wright, B. D. (1988). Practical adaptive testing. In: Rasch measurement transactions. Rasch Measurement SIG.

Future Development: Towards Semantic Compliance Checking

Ildikó Szabó

1 Ontology Matching

The tendency to create a transparent, semantically searchable web boosts the need to combine these ontologies by merging, transforming, integrating, aligning, versioning and mapping them. The methods in this field are distinguished on the basis of their goals.

Ontology management uses the ontology library system to store, align and maintain ontologies (Fensel et al. 2002).

Ontology mediation is aimed at reusing ontologies throughout various heterogeneous applications by determining and overcoming differences between ontologies. This domain consists of three areas: "ontology mapping, which is mostly concerned with the representation of correspondences between ontologies; ontology alignment, which is concerned with the (semi-)automatic discovery of correspondences between ontologies; and ontology merging, which is concerned with creating a single new ontology, based on a number of source ontologies" (Bruijn et al. 2006, p. 1).

Ontology mapping defined by Su (2002, p. 3) means that "for each concept (node) in ontology A, try to find a corresponding concept (node), which has the same or similar semantics, in ontology B and vice versa."

Ehrig and Sure (2004, p. 5) determined the map : Oi1 $\rightarrow$ Oi2 ontology mapping function on ontologies defined as O := $(C, H_C, R_C, H_R, I, R_I, A)$[1] tuples in the following way:

---

[1] An ontology O consist of concepts arranged into a subsumption hierarchy HC, relations RC between them, properties (specific relations) arranged into hierarchy HR, instances I as specific concepts and relations RI between them and finally additional axioms to infer knowledge from an already existing one.

I. Szabó (✉)
Corvinno Technology Transfer Center, Budapest, Hungary
e-mail: iszabo@corvinno.hu

© Springer International Publishing Switzerland 2016
A. Gábor, A. Kő (eds.), *Corporate Knowledge Discovery and Organizational Learning*, Knowledge Management and Organizational Learning 2,
DOI 10.1007/978-3-319-28917-5_7

155

map($e_{i_1j_1}$) = $e_{i_2j_2}$, if sim($e_{i_1j_1}$, $e_{i_2j_2}$) > t threshold, then $e_{i_1j_1}$ is mapped onto ei_2j_2. This means they are semantically identical. Each entity $e_{i_1j_1}$ is mapped to at most one entity $e_{i_2j_2}$, where sim($e_{i_1j_1}$, $e_{i_2j_2}$) measures similarity between two entities, where $_{i_1} \neq _{i_2}$, and sim(x,y) is a similarity function [e.g. sigmoid function (Ehrig and Sure 2004, p. 10)]. The range of this function is [0;1]. It is a reflexive, symmetric function. The triangular in equation is valid for it, if the in equation is true. This is the inverse of distance. If two objects are identical, their value is 1. e_{ij} is entities of Oi with $e_{ij} \in \{C_i, R_i, I_i\}$, $j \in N$.

These ontology mapping definitions use elements of set theory. Kalfoglou and Schorlemmer (2003, p. 3) defined ontology mapping based on logical theories. They considered ontologies as "a pair O = (S,A), where S is the (ontological) signature, and A is a set of (ontological) axioms—specifying the intended interpretation of the vocabulary in some domain of discourse." An ontological signature is a mathematical modelling tool to describe ontologies with a hierarchy of concept or class symbols modelled as a partial ordered set. Morphisms (structure-preserving mappings between mathematical structures) are used to define total ontology mapping from O1 = (S1,A1) to O2 = (S2,A2).

Consequently, ontology mapping is responsible for creating candidate mappings from concepts of different ontologies that are an appropriate base for analyzing them according to ontology matchers.

Ontologies are born from different perspectives used for scrutinizing a given domain. Conceptualization, application of paradigms or concept description, terminological discrepancies (usage of synonyms, homonyms etc.) explain the diversity of these ontologies.

Depending on the depth of conceptualization variant level of expressivity—including syntax, logical representations, and semantics of primitives—are required to formalize these ontologies. Hence ontology mismatches are also experienced at language or at an ontology level (Klein 2001).

At ontology level, discrepancies are derived from different name conventions, specification of the ontologies and dissimilarity in their connections between ontology elements. So semantic and syntactic scrutinizing are needed to discover these types of discrepancies at an element and structural level. Euzenat and Shvaiko (2013, p. 65) elaborated this categorization to classify different matching techniques (Fig. 1).

Element-level matching techniques concern only matching entities and its instances without any information about their relationships with other entities or instances. These techniques are distinguished by the basis of this comparison that can be extracted from the names or name descriptions of these entities as strings (string-based techniques) or words (language-based techniques) or from the definitions of these entities according to types, cardinality of attributes and keys. Usage of external sources—linguistic resources, history of previously matched ontologies, and upper level ontologies can enrich these techniques.

Element level techniques use different similarity measures to estimate the proximity of entities or instances. These measures are used in the ontology learning

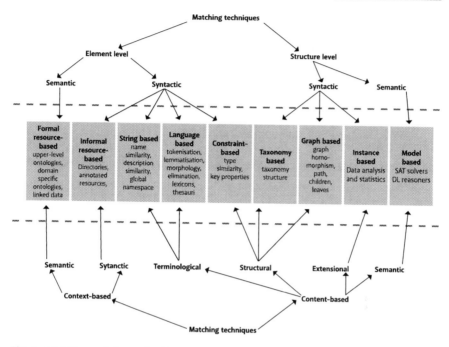

Fig. 1 Matching techniques classification by Euzenat and Shvaiko (Otero-Cerdeira et al. 2015, p. 955)

and mapping field that were elaborated upon in Chapter "ProMine: A text mining solution for concept extraction and filtering".

The structure is composed of relations between ontology entities and their instances that appeared in ontology. The shapes of these structures are examined using *structure-level matching techniques* that are distinguished based on the representation form of these structures. Graph-based technique regards both ontologies as labelled graphs and investigate similarity comparisons between their nodes based on the positions of these nodes within their own graph.

But semantic business process management also uses these methods to analyze business processes for executing optimization, compliance checking (Gábor et al. 2013). During the next section, the applicability of general ontology matchers will be investigated to use them for the purpose of SBPM.

2   Evaluating Ontology Matching Tools

The evaluation criteria are determined by the subject and goal of matching processes. In our use cases, process ontologies are the subjects and discovering structural discrepancies are the goals. Nevertheless, the appropriate ontology matching tool should be connected to tools developed in the ProKEX. The following functional requirements were chosen as criteria:

Integrality

- reusability: the deliverables of the ProKEX project were implemented mostly in JAVA, so the question is does the ontology matcher have any available source code written in JAVA. Is it freely modifiable or just under the aegis of a license agreement.
- report type: the goal is to inform stakeholders about the compliance of their process using guidelines, best practice etc. Different tools create technical report first of all, hence these reports must be restructured based on the requirements of these stakeholders. The best format of this technical report is text, not a visualized one.

Dynamic Processing

- human intervention: which activities of the matching process require human intervention
- automatic data load: the process ontologies are transformed from process models, so it is necessary to load these data in an automatic manner to manage changes that have come up in the ontology
- ontology format: process ontologies are interpreted in XML, RDF or OWL. It is necessary to examine which formats are managed by the matcher.

The type of matching method: business process models have a unique structure, hence it must be examined according to the tool, thus the applicability of the structure-level matching technique is required.

A literature review presented by Otero-Cerdeira et al. (2015) investigated the evolution of 60 different matching tools based on their presence progress in the literature. Several tools from them were developed for the Ontology Alignment Evolution Initiative contest. AgreementMaker, LogMap, Rimom, and Yam++ have been presenting in journals for at least three years.

We aimed at developing a system that is capable of processing documents to extract process ontology from them and use this ontology to investigate actual business processes presented by another process ontology. However, Agreement-MakerLight is specialized for biomedical ontology, and Rimom-Im is an iterative instance matching tool (Shao et al. 2014), and our purpose is to check process ontology classes in the general field. Hence only LogMap, Yam++ were selected for the examination. But the Protégé ontology development editor is a popular, open source tool with a built-in ontology matching function, hence it was added to them.

2.1   LogMAP2

The predecessor of this ontology matcher was developed for the OAEI competition. It manages not only the OWL 2 format, but also that of the RDF/XML, OWL/XML, OWL Functional, KRSS and Turtle as well. A standalone version and an SVN repository ensure access to this tool. Its first stage is to collect concepts matching lexically by investigating their labels and URIs, and in the meantime the recall is maximized. Its second stage is to assess these candidate mappings based on the high similarity values of the mapped elements and if there is also a candidate mapping between their neighbors. Non-reliable mappings are discarded based on scrutinizing propositional encoding of these mappings, using reasoning-based algorithms, semantic indexes and user interaction (mostly in the interactive mode). The tool creates three OWL files for presenting the mapping results (Jiménez-Ruiz et al. 2012).

2.2 Yam++

This tool is importable into Eclipse from existing projects. The matching process is customizable by the user and executed at three levels:

- at element level the main goal is to collect candidate mappings
- at structural level the algorithm discard the unreliable mappings based on the structures of the input ontologies
- at semantic level it optimizes the set of remaining candidate mappings by using a global constraint optimization method proposed in Alcomox tool.

It applies either to machine learning models or information retrieval methods that depend on the existence of a gold standard dataset on which the precision of the algorithm can be calculated. This tool is applicable for multi-language ontologies as well. The user can set this tool by choosing similarity metrics, a gold standard dataset or a machine learning model (Ngo and Bellahsene 2012).

2.3 Protégé 4 OWL Diff

This tool used some procedures of Prompt (Noy et al. 2015) loosely. It is runnable by following the Tools- > Compare Ontologies path within the Protégé 4.X or 5.0 ontology development environment. Its input sources are the same that Protégé deals with the same formats as LogMap2 does. The purpose of this tool is to discover changes made during updating the original version of an ontology, hence it compares the source (original) ontology with the target (updated) ontology. It therefore focuses on structural alignment that chiefly means axiomatic investigation of the candidate entities. Entities are aligned with each other if:

- they have the same IRI or differentiate only in namespace,
- they have the same annotations regarding as rendering,
- they have a relationship with the same parent and child (sometimes this is questionable),
- they have the same siblings including the misspelled ones or the changed ones from singular to plural.

Having collected the alignments during the alignment phase, a technical and a visualized report are created during the explanation phase. The mappings will be sorted into four blocks:

- an entity in the Created block appears in the target ontology but there is no analogue in the source ontology.
- an entity in the Deleted block appears in the source ontology but there is no analogue in the target ontology.
- an entity in the Modified block has the same name in both ontologies but some of their axioms are different.
- an entity in the Renamed block that appears in the target ontology corresponds to an entity in the source ontology, but the names of the two entities are different.

The source code is written in Java and available through an SVN repository, hence it can be built in a Java-based OWL API to load any input formats—OWL-DL, RDF, OWL 2—handled by Protégé and transform the technical report into a form that can be interpreted by stakeholders (Redmond and Noy 2011) (Table 1).

Source codes of these tools are freely available, hence they can be connected to the tools elaborated in the ProKEX by developing a JAVA API. But the main advantages of Protégé 4 OWL Diff are that it provides a text-based report that can be converted into a meaningful, transparent report for process owners and it chiefly investigates at a structural level that provides important information about the structure of business processes. Its weakness is the slightly element-level investigation that can be resolved by using similarity measures. This is the reason why this tool was chosen for implementing the ontology matching part of the next system.

Table 1 The comparison of ontology matchers

		LogMAP	Yam++	Protégé tool
Integrality	Reusability	Standalone version and SVN repository	Importable JAVA project	SVN repository
	Report type	OWL files	RDF file and screen	Text, Java applet
Dynamic processing	Human intervention	Only in the interactive mode	For configuring the tool	None
	Automatic data load	Through a JAVA API	Through a JAVA API	Through an OWL API
	Ontology format	OWL 2, OWL/XML, RDF/XML etc.	OWL	OWL 2, OWL/XML, RDF/XML etc.
Matching method		Mostly element-level, but using structural investigation for discarding mappings	Mostly element-level, but using structural investigation for discarding mappings, and semantic level	Mostly structure-level

3   Process Ontology Building and Matching

The next figure presents how the process ontology building and matching system
works (Fig. 2).

The process has two parallel branches:

- one of them creates a process ontology from business process models through
  XSLT conversion (see in Chapter "Corporate semantic business process management"). The main classes correspond to the main building blocks of a process
  model e.g. roles, process step, I/O system etc. see Figure 3.
- the other branch uses this structure and the previous process ontology to build
  the reference process ontology.

The output of these branches are fed with the chosen ontology matcher to create
a report for discovering dissimilarities between these processes.

There are two ways that can be considered for building the reference process
ontology:

- without any prior information about the actual business process (general
  procedure)
- using process elements of the actual business process model as a comparison
  base for identifying the dissimilarities from this while the reference document is
  processing (heuristic procedure).

Elaborating the general procedure is very challenging, but more difficult than the
creation of a heuristic one. Hence, we initially started to identify the stages of a
heuristic procedure the outcomes of which would be used for creating general
algorithms in the future.

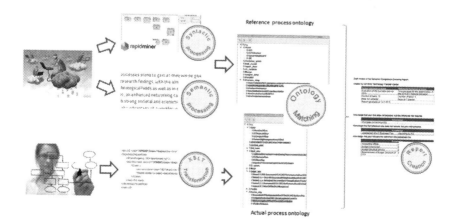

Fig. 2 Process ontology building and matching

Fig. 3 The main classes of a process ontology

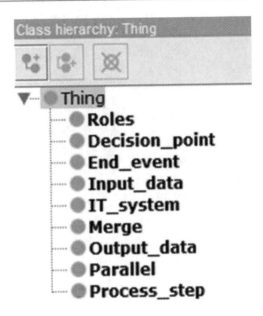

3.1   Heuristic Procedure for Building Reference Ontology

The input sources of this procedure are reference documents—like best practices, standards, protocols, regulations etc.—in unstructured or poorly structured formats (e.g. doc, xhtml etc.) and the basic structure of the process ontologies defined by Adonis building blocks (see Fig. 3.).

The procedure extracts the process stages from the actual process ontology and tries to identify them one after the other in the reference document. Grammatical and semantic rules will be combined to reach a higher level of precision. Process stages are generally formalized by verb and noun combinations. The Rapidminer process can facilitate the collection of verbs and nouns from a given text (Fig. 4).

We can use this type of combination for dividing the text into task or activity-related parts. But now future development means completing the semantic rules with these kinds of grammatical rules.

The goal of semantic rules is to identify the building blocks of process models (process steps, roles etc.) based on their meaning. We formalized rules for extracting related expressions from sentences. For example 'by the' and 'role of' expressions split a sentence into two parts. The right side of them is likely a role.

Input data and output data can be identified by '* document', '* material', '* model' in that * replaces one or more words.

The processing procedure has the following stages:

- collecting terms from a given process stage of the actual process ontology
- processing each sentence of the reference document to find all the terms or its subsets within them

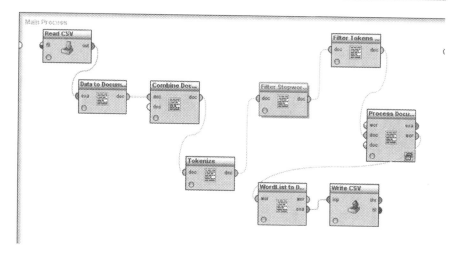

Fig. 4 Rapidminer process for extracting part-of-speech tags

- if a sentence contains enough words (parameterized with a threshold), the algorithm will accept this sentence as similar to the given process stage. It will insert it or its relevant part into the reference process ontology as a subclass of the 'Process_step' class.
- having discovered this likely relevant sentence, the other building blocks (such as roles, input data, output data etc.) are discovered around it (within a given circle). The identified expressions will be inserted into the reference ontology as a subclass of the related main class (Role, Input_data etc.) and they will be connected to the new process stage too.

Having created the reference process ontology, Protégé 4 OWL Diff ontology matcher aligns the ontologies to each other. This tool creates a technical report split into the four above-mentioned blocks.

Created blocks contain information about process elements in the reference process ontology, but not in the actual process ontology.

Deleted blocks provide information about process elements in the actual process ontology, but not in the reference process ontology.

Renamed and Modified blocks contain information about correspondent process elements with name or axiomatic differences.

This technical report is not too perspicuous for a process owner, hence the information of these blocks are transformed into more transparent and sectioned report based on the information need of process owners.

This contains the following elements:

- name and task number of the process,
- missing process elements from the actual business process for working as required,
- unnecessary process elements in the actual business process for working as required,
- common, slightly modified process elements for considering their improvement.

A way to extend this solution is to merge process ontologies with domain ontology containing knowledge for executing these process. Hence the related knowledge elements are connected to the process stages with a 'required by' relation. These relations are extracted from the concept groups created during the job role-task assignment (see in Chapter "Ontology tailoring for job role knowledge"). Domain ontologies and concept groups are stored in the Studio system (details are in Chapter "STUDIO: Ontology-centric knowledge-based system"). Java Webservice is responsible for importing the concept groups and domain ontology from the Studio system, and Protégé function can help to merge them with the actual process ontology.

Processes do not require all of the knowledge elements stored in the domain ontology, since the domain ontology will be tailored based on either the requirements of the actual business process or the reference business process. It means that an actualized and a specialized version will be created from the domain ontology. The actualized version can be extracted from the concept groups. The accepted knowledge elements will receive a data type attribute whose value is 'actual' in the actual merged process ontology, and 'ref' in the reference merged process ontology. Ontology matching focuses on discovering missing, unnecessary and common knowledge elements regarding the whole processes generally on the one hand or role-specific knowledge elements on the other hand. Revealing role-specific knowledge elements requires the filtering of knowledge elements related to the specific role. DL Query provides a formula for querying ontologies semantically, hence it is an appropriate instrument for executing this task (Table 2).

This tool was written in Java, using OWL API, DLQueryExample.java and SVN repository of Compare Ontologies function. Having filtered the ontologies, the algorithm of Protégé 4 OWL Diff runs, and based on its technical report the above-mentioned report will be created, but referencing the role.

The applicability of this tool will be presented by the evaluation process of the European Institute of Innovation and Technology in the next sections.

Table 2 DL Queries for filtering the knowledge elements

Description	DL Query
Knowledge possessed by a given role	Required_by some (performed_by only <Role>)
Knowledge required to execute the process	Required_by some (belongs_to only <Process>)
Knowledge required by a given task	Required_by some (<Process_step>)

3.2 Usability of This Ontology-Based Querying and Matching Tool in SBPM

Semantic business process management (see in Chapter "Corporate semantic business process management") also has a role in process improvement and compliance checking. Combining semantics with business process management add machine readable meaning to these processes.

The ontology approach ensures a unified view of business processes and a tool for querying and matching them semantically. Investigating actual business process in the light of reference business process raises questions about compliance of whole actual processes or provides an opportunity to improve enterprise knowledge assets for running actual business process faster. The above-mentioned ontology-based querying and matching tool is capable of answering the following questions potentially raised by process owners:

Questions about the compliance of the whole process:

- To what degree are the various process elements (process stages, roles, I/O data) from the actual process overlapped by the similar ones from the reference business process?
- Do the actual and reference business processes follow the same process flow?
- Do the actual business processes satisfy certain explicit segregation of duty criteria?

Questions to aim at improving the enterprise knowledge asset:

- To what degree are the roles in actual processes overlapped by the roles in reference processes regarding knowledge required by task which is performed by a given role?
- Is the same knowledge required for executing the same tasks in the actual and the reference process?
- Is the same knowledge required for holding down the same job roles in the actual and the reference process or process group?

These questions imply a layered query, initially at task level, then at process level, and finally at process group level. In this way, a detailed report is produced based on the knowledge required to execute the processes.

Answering the first question groups require reference process ontologies built from reference documents (e.g. best practices, standards, protocols etc.). An elementary process ontology building procedure was presented in the previous section. This will be illustrated by using the evaluation process of the European Institute of Innovation and Technology. The next stage to provide meaningful answers is to match the reviewed reference process ontology with the actual process ontology. The matching report on the discrepancies and similarities between the reference and actual process ontology will be presented in the subsequent section.

Providing information related to the last question groups need to merge each process ontologies with the actualized or specialized version of the domain ontology. Having created each merged version and having filtered them with the required role, a knowledge-based ontology matching will be executed for answering these questions. Examples on the fund management domain will be showed as a use case related to this building and matching system.

4 Process Ontology Building on the Fund Management Field

This illustration uses the evaluation of KIC's past performance process (see Chapter "Corporate knowledge discovery and organizational learning: The role, importance, and application of semantic business process management—The ProKEX case"). The actual business process was transformed into the actual process ontology through XSLT transformation (Fig. 5.).

The compliance of this process will be investigated in the next section, but first of all the process ontology building procedure was applied to create the reference process ontology. In the first stage, the roles were identified by using 'by the' semantic rule. 'Governing Board' and 'IT Director' were extracted by this rule. In the second stage, the words of the process stages from the actual business process were used to identify likely related process stages in the reference document. E.g. the 'Selection of External Experts' process stage was split into 'selection', 'of', 'external', 'experts' terms and the expression 'External Independent Experts' was identified by them. The term 'Governing Board' was found within a given radius, hence it was connected to this process step in the reference process ontology (Fig. 6).

Having refined this algorithm by using similarity measures (see Chapter "ProMine: A text mining solution for concept extraction and filtering"), the classes in the reference process ontology can be renamed as classes with high similarity values of the actual process ontology. Hence 'ExpertsSelected' and 'ExtErnalIndependentExpErtSelected' classes were merged into 'SelectionOfExternalExperts' class. This conversion is needed because the structural investigation of the Protégé 4 OWL Diff tool must be completed with an element-level investigation.

4.1 Ontology-Based Compliance Checking on a Fund Management Process

We can examine the compliance of the actual business processes regarding the reference business process by using the Protégé 4 OWL Diff ontology matcher. After running this matcher the next report was produced in text format.

A created block shows process elements that are in the reference business process represented by the reference process ontology, but not in the actual business

Fig. 5 The evaluation process of EIT in Protégé 5

process represented by the actual process ontology. The next figure shows that these process steps were identified in the reference document that can be connected to the actual process steps. These can be required process stages, but actual ones as well. In the latter case the algorithm should be improved. This technical report shows that two roles—EIT Director and Governing Board—which were extracted from the texts (Fig. 7).

After the technical report was processed it was revealed that the EIT Director role is also part of the actual process ontology. This result could also be presented in table shape (Fig. 8).

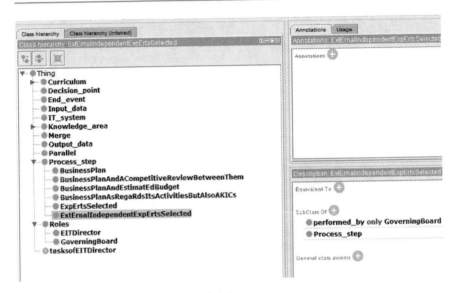

Fig. 6 Reference process ontology in Protégé 5

```
Created BusinessPlan
-------------------------------------------------------------
Added:    Class: BusinessPlan
Added:    BusinessPlan SubClassOf Process_step
Added:    BusinessPlan SubClassOf performed_by only GoverningBoard
-------------------------------------------------------------
Created BusinessPlanAndACompetitiveReviewBetweenThem
-------------------------------------------------------------
Added:    Class: BusinessPlanAndACompetitiveReviewBetweenThem
Added:    BusinessPlanAndACompetitiveReviewBetweenThem SubClassOf Process_step
Added:    BusinessPlanAndACompetitiveReviewBetweenThem SubClassOf performed_by only EITDirector
-------------------------------------------------------------
Created BusinessPlanAndEstimatEdBudget
-------------------------------------------------------------
Added:    Class: BusinessPlanAndEstimatEdBudget
Added:    BusinessPlanAndEstimatEdBudget SubClassOf Process_step
Added:    BusinessPlanAndEstimatEdBudget SubClassOf performed_by only EITDirector
-------------------------------------------------------------
Created BusinessPlanAsRegaRdsItsActivitiesButAlsoAKICs
-------------------------------------------------------------
Added:    Class: BusinessPlanAsRegaRdsItsActivitiesButAlsoAKICs
Added:    BusinessPlanAsRegaRdsItsActivitiesButAlsoAKICs SubClassOf Process_step
Added:    BusinessPlanAsRegaRdsItsActivitiesButAlsoAKICs SubClassOf performed_by only GoverningBoard
-------------------------------------------------------------
```

Fig. 7 Required process stages extracted from the reference document

Description	Baseline Axiom	New Axiom
Renamed	Class: EITDirector	Class: EITDirector

Enties with a common rendering are matched.

Fig. 8 Common role in both business processes

Description	Baseline Axiom	New Axiom
Renamed	Class: SelectionOfExternalExperts	Class: SelectionOfExternalExperts
Added		SelectionOfExternalExperts SubClassOf performed_by only GoverningBoard
Deleted	SelectionOfExternalExperts SubClassOf accountable_role only PartnershipManagementHeadOfUnit	
Deleted	SelectionOfExternalExperts SubClassOf cooperative_role only EducationOfficer	
Deleted	SelectionOfExternalExperts SubClassOf followed_by only ContractingOfExperts	
Deleted	SelectionOfExternalExperts SubClassOf performed_by only KICProjectOfficer	

Enties with a common rendering are matched.

Fig. 9 Discrepancies in the common process stage

The 'Selection of External Experts' process stage was identified as a common process stage in the previous phase. The ontology matching tool discovered the discrepancies between them. The following picture shows that three different roles—accountable role, cooperative role and executive role—are distinguished in the actual business process, but only one—the executive role—in the reference business process. Governing board is not a role, but a board, hence the process ontology building algorithm should be improved in order to eliminate these types of executors during the extraction. We can introduce, for example, semantic stop word lists that are connected to a specified ontology elements. But this may also show that the reference document is too general for building a precise reference process ontology (Fig. 9).

As we have seen, the ontology matching report can contain some useful information about the structure of the actual business process in respect to the reference business process. The next section will illustrate that these kinds of reports are convertible into versions that can be more easily interpreted by process owners, and in the meantime the goal of those reports are to facilitate control over the actual business process.

4.2 Knowledge-Based Business Process Control on Fund Management Standards

Business processes require knowledge to run them. This knowledge builds into human minds. Ontology tailoring and job role-task alignment in the ProKEX project are able to extract this knowledge from business processes. During the alignment of business processes to the relevant domain ontology, concept groups are created in the Studio system. These concept groups contain information about the mapping between tasks and required knowledge by them. Having imported these concept groups and merged them with the actual process ontology, a merged ontology is created that is applicable for investigating a process control such as the segregation of duties based on knowledge. A similar merged ontology can be evolved on the reference side if this tool is completed by using not only business processes as input source but also texts like reference documents.

The main task is to investigate if the same role requires the same knowledge base on actual and reference sides or certain knowledge is required by only one role from them that can entail executing a more thorough investigation. The overlapping of certain job roles must be analyzed in order to discover the level of satisfaction in respect to the segregation of duties control or the knowledge base of employees with the given job role should be enhanced to meet the reference requirements.

The first task was to filter the actual and reference process ontology by a given job role. DL Query helped to execute this task. In the actual business process, the EIT Director has an accountable role, meanwhile it has an executive role in the reference business process, hence different DL Queries were used to filter the ontologies. Knowledge and tasks are related to each other through 'required by' relation, hence 'Required_by some (accountable_role only (EITDirector))' query was run on the actual process ontology and 'Required_by some (performed_by only (EITDirector)) on the reference process ontology (Fig. 10).

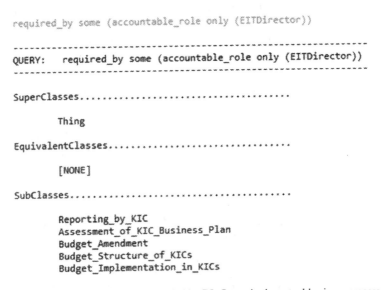

Fig. 10 First-level knowledge areas identified by DL Query in the actual business process

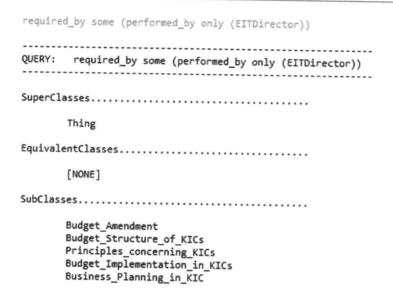

```
required_by some (performed_by only (EITDirector))

------------------------------------------------------------
QUERY:   required_by some (performed_by only (EITDirector))
------------------------------------------------------------

SuperClasses.....................................

        Thing

EquivalentClasses................................

        [NONE]

SubClasses.......................................

        Budget_Amendment
        Budget_Structure_of_KICs
        Principles_concerning_KICs
        Budget_Implementation_in_KICs
        Business_Planning_in_KIC
```

Fig. 11 First-level knowledge areas identified by DL Query in the reference business process

The identified knowledge elements, its sub knowledge areas and other knowledge areas connected to them through 'required_knowledge_of' relation were stamped by a datatype. The value of this datatype was 'actual' in the actual process ontology, and 'ref' in the reference process ontology (Fig. 11).

The ontology matcher runs on these identified knowledge elements and the next report was created from the technical report mentioned in the previous section. This report establishes which knowledge elements are not important to execute the process based on the reference, which knowledge elements are important but lacking or existing now. The process owner can use this report for executing a more thorough investigation or improving the enterprise knowledge asset of employees (Fig. 12).

Our approach provides process owners with a solution for investigating business processes at the structural level and at the level of knowledge-related activities. In this way, improving and controlling business will be more sophisticated, reflecting the hidden knowledge behind the processes.

Draft version of the Semantic Compliance Checking Report

created by Corvinno Technology Transfer Center

Actual business process	Reference business process
Evaluation of the business plan for 2014	The principles for the determination of the 2016 EIT financial allocation
Number of tasks: 10	Number of tasks: 5
Role: EIT Director	Role: EIT Director
Report generated at 14-7-2015	

Knowledge that your role does not possess, but the reference role requires

Knowledge	
Principles concerning KICs	

Knowledge that the reference role does not require, but your role possess

Knowledge	
Assessment of KIC Business Plan	Reporting by KIC

Knowledge that your role and the reference role possesses too

Knowledge	
Accounting Officer	Authorising Officer
Budget Amendment	Budget Implementation in KICs
Budget Structure of KICs	Business Planning in KIC
Establishment of Budget Structure of KICs	Expenditure Operations in KICs

Fig. 12 Compliance checking report

5   Conclusion

This chapter presented a solution for improving or controlling business processes
that can integrate the various tools developed in the ProKEX project. XSLT
conversion creates the actual process ontology from the actual business process.
Similarity measures used in the text mining tool can be used to refine the reference
process ontology extracted from the reference documents. Ontology tailoring and
job role-task assignments facilitate the alignment of the knowledge to the job role
required by them and create tailored—actualized and specialized—domain
ontologies related to the actual and reference business processes, using concept
groups. These tools can be the main building blocks of the above-mentioned
process ontology building and matching system the framework of which has
already been implemented using the OWL API, DLQueryExample.java and SVN
repository of Protégé 4 OWL Diff ontology matcher.

The algorithm will be substantially refined in the future. Its precision will be measured by using evaluation methods such as the confusion matrix, ROC curve etc.

References

Bruijn, J., Ehrig, M., Feier, C., Martíns-Recuerda, F., Scharffe, F., & Weiten, M. (2006). Ontology mediation, merging, and aligning. In J. Davies, R. Studer, & P. Warren (Eds.), *Semantic web technologies: Trends and research in ontology-based systems* (pp. 95–113). Chichester: Wiley.

Ehrig, M., & Sure, Y. (2004). Ontology mapping—An integrated approach. In C. J. Bussler, J. Davies, D. Fensel, & R. Studer (Eds.), *The Semantic web: Research and applications* (pp. 76–91). Berlin: Springer.

Euzenat, J., & Shvaiko, P. (2013). Classifications of ontology matching techniques. In *Ontology matching* (pp. 73–84). Berlin: Springer.

Fensel, D., Davies, J., & Van Harmelen, F. (2002). *Towards the semantic web: Ontology-driven knowledge management*. Chichester: Wiley.

Gábor, A., Kő, A., Szabó, I., Ternai, K., & Varga, K. (2013). Compliance check in semantic business process management. In Y. T. Demey & H. Panetto (Eds.), *Lecture notes in computer science* (Vol. 8186, pp. 353–362). Berlin: Springer.

Jiménez-Ruiz, E., Grau, B. C., Zhou, Y., & Horrocks, I. (2012). *Large-scale interactive ontology matching: Algorithms and implementation*. ECAI, pp. 444–449.

Kalfoglou, Y., & Schorlemmer, M. (2003). Ontology mapping: The state of the art. *The Knowledge Engineering Review, 18*, 1–31. doi:10.1017/S0269888903000651.

Klein, M. (2001). *Combining and relating ontologies: An analysis of problems and solutions.* Workshop on ontologies and information sharing, IJCAI, pp. 53–62.

Ngo, D., & Bellahsene, Z. (2012). YAM++: A multi-strategy based approach for ontology matching task. In A. ten Teije, J. Völker, S. Handschuh, H. Stuckenschmidt, M. d'Acquin, A. Nikolov, N. Aussenac-Gilles, & N. Hernandez (Eds.), *Knowledge engineering and knowledge management* (pp. 421–425). Berlin: Springer.

Noy, N., Klein, M., Kunnatur, S., Chugh, A., & Falconer, S. (2015). PROMPT—Protege Wiki. Accessed July 14, 2015, from http://protegewiki.stanford.edu/wiki/PROMPT

Otero-Cerdeira, L., Rodríguez-Martínez, F. J., & Gómez-Rodríguez, A. (2015). Ontology matching: A literature review. *Expert Systems with Applications, 42*, 949–971. doi:10.1016/j.eswa.2014.08.032.

Redmond, T., & Noy, N. (2011). *Computing the changes between ontologies*. Joint Workshop on Knowledge Evolution and Ontology Dynamics, pp. 1–14.

Shao, C., Hu, L., & Li, J. (2014). RiMOM-IM results for OAEI 2014. In P. Shvaiko, J. Euzenat, M. Mao, E. Jiménez-Ruiz, J. Li, & A. Ngonga (Eds.), *Proceedings of the 9th International Workshop on Ontology Matching* (pp. 149–154).

Su, X. (2002). *A text categorization perspective for ontology mapping*. Norway: Department of Computer and Information Science, Norwegian University of Science and Technology.